Samuel Wainwright

Scientific sophisms

a review of current theories concerning atoms, apes, and men

Samuel Wainwright

Scientific sophisms

a review of current theories concerning atoms, apes, and men

ISBN/EAN: 9783337414450

Printed in Europe, USA, Canada, Australia, Japan

Cover: Foto ©berggeist007 / pixelio.de

More available books at **www.hansebooks.com**

SCIENTIFIC SOPHISMS.

*A REVIEW OF CURRENT THEORIES
CONCERNING ATOMS, APES, AND MEN.*

BY

SAMUEL WAINWRIGHT, D.D.,
AUTHOR OF
"CHRISTIAN CERTAINTY," "THE MODERN AVERNUS," ETC.

NEW YORK:
FUNK & WAGNALLS,
10 AND 12 DEY STREET,
1883.

SCIENTIFIC SOPHISMS.

CONTENTS.

	PAGE
ANALYTICAL OUTLINE	vii

I.
THE RIGHT OF SEARCH. 1

II.
EVOLUTION 11

III.
"A PUERILE HYPOTHESIS" 33

IV.
"SCIENTIFIC LEVITY" 45

V.
A HOUSE OF CARDS 61

VI.
SOPHISMS 75

VII.
PROTOPLASM 99

VIII.
THE THREE BEGINNINGS 149

IX.
THE THREE BARRIERS 169

X.
ATOMS 187

XI.
APES 203

XII.
MEN 225

XIII.
ANIMA MUNDI 251

APPENDIX. 299

ANALYTICAL OUTLINE OF CONTENTS.

CHAPTER I.
THE RIGHT OF SEARCH.

AGNOSTICISM : and
Gnosticism :
    Its Pretensions.
Prof. Clifford—
    His "Ethics of Religion";
    His new divinity.
Prof. Tyndall—
    His assumptions ;
    His admissions.
Their relation to
MATERIALISTIC ATHEISM :
    Is it true ?
    Is it demonstrable ?
    Is it Scientific ?

CHAPTER II.
EVOLUTION.

EVOLUTION :
    Theories of :
    Three main varieties :
        The Theistic,
        The Atheistic,
        The Agnostic.
Their relation to the doctrine of

viii *Analytical Outline of Contents.*

Development :
    Mr. Darwin's "view"; and his Opinion.
        His "opinion" may be questioned ; and
        His "view" has not been shown to be true.
    Is strongly Theistic,
    Is shown by Professor Mivart to be
        "Not *The* Origin of Species," and
        "Not antagonistic to Christianity."
The Theistic Doctrine of Evolution :
        (Its three main Varieties)
    Maintained by Mr. Darwin ; but
    Opposed by Professors Huxley and Tyndall.
    Prof. Tyndall "abandons," once for all "the conception of creative acts."
    Prof. Huxley excludes "the intervention of any but what are termed secondary causes."
Evolution :
    As strictly defined,
    As popularly understood.
    The validity of the Facts
    Independent of every Theory as to their Cause.
    The Phenomenal Sequence,
        Not the Ideal Hypothesis,
    A Universal Law.
The Ideal Hypothesis, which
        "Derives man in his totality from the interaction of organism and environment through countless ages past."

CHAPTER III.

"A PUERILE HYPOTHESIS."

Evolution :
    "Baldest of all philosophies"
    Involves two points.
I. ASCENSIVE DEVELOPMENT :
    Negatived by
        "The positively ascertained truths of Palæontology."

II. THE TRANSMUTATION OF SPECIES.
 "Not Proven" (Prof. Huxley).
 "Of direct and positive testimony"
 "There is no fragment whatever" (Dr. Elam).
 Mr. Darwin's admissions
  "Fatal" to his theory
 Condemned by Prof. Mivart.

CHAPTER IV.

"SCIENTIFIC LEVITY."

AGNOSTIC EVOLUTION:
 An Unverified Hypothesis
 Based on two subordinate hypotheses
  Equally unverified.
 (1) Spontaneous Generation.
 (2) The Transmutation of Species.
SPONTANEOUS GENERATION.
 "Does life grow out of dead matter?" (Prof. Whewell.)
 "It is a result absolutely inconceivable." (Mr. Darwin.)
 "Not supported by any evidence." (Dr. Carpenter.)
 "Scientific Levity." (Humboldt.)
 From Matter to Life:
  The attempts to bridge the chasm
   Have all failed.
  The "nucleated vesicle"
   Is on the wrong side of the gulf.
  The "chemico-electric operation"
   Is a mere "supposition."
  The "Protogenes of Haeckel," and
   Dr Elam's refutation of Mr. Spencer.
  The "line of demarcation
   between the organic and the inorganic
    Is as wide as ever."
Chemistry: Its century of triumphs.
 Its one conspicuous Failure. Hence

SPONTANEOUS GENERATION is
"An astounding hypothesis" (Dr. Carpenter)
"Vitiated by error" (Prof. Tyndall), and
"Utterly discredited." (Virchow.)

CHAPTER V.

A HOUSE OF CARDS.

Agnostic Evolution : Not scientifically true.
"A flimsy framework of hypotheses." (Dr. Elam.)
Devoid of "experimental demonstration." (Tyndall.)
Its Fundamental Proposition :
Condemned by Scientific Authorities
"The older and honoured chiefs in Natural Science ;"
(Darwin.)
"A minority of minds of high calibre and culture."
(Tyndall.)
The New Syllogisms :
"Probable" ; "provisional" ; "uncertain."
"Reason to suppose :" (Mr. Spencer)
"I can imagine ;" (Prof. Tyndall)
"It is conceivable." (Mr. Darwin)

CHAPTER VI.

SOPHISMS.

I. Prof. Haeckel's Genealogy :
    Its hypothetical completeness : Dependent on
    Its Continuity—"in nubibus."
    Refuted by Du Bois Reymond.
His Fundamental Postulates :
    Incapable of Proof.
    Monera ; Gastreada ; Amphioxus.
Accepted by Mr. Huxley. And yet
    Mr. Huxley admits that
    The doctrine of Evolution involves the assumption of
    Spontaneous Generation ; while this last has
"No experimental evidence in its favour."
    Supported by "no valid or intelligible reason."

Analytical Outline of Contents. xi

II. BIOGENESIS:
    Harvey, and Francesco Redi.
    Paradoxical position of Mr. Huxley.
    (1) As a Biogenist, he holds that
        "All living matter has sprung from pre-existing living matter."
    (2) As an Abiogenist, he thinks that
        Life may "some day be artificially brought together."
    (3) He thinks this has never yet been done. But yet
    (4) If he had been living in the remote Past
        He should expect to have seen it done.

III. Prof. Tyndall's Fallacies
    (1) The "impulse inherent in primeval man."
    (2) "The possible play of molecules in a cooling planet."
    (3) "Physical theories beyond the pale of experience."
    (4) His imagining the unimaginable.
        (*a*) The passage from physics to consciousness
            Is "unthinkable." And yet he says
        (*b*) "By an intellectual necessity
            I cross the boundary."
    (5) He tells us of
        (*a*) "The chasm between the two classes of phenomena."
        (*b*) He declares this chasm to be "Intellectually impassable"; and yet
        (*c*) He proclaims his belief in "The Continuity of Nature."
    (6) The Continuity of an "impassable chasm"
        (*a*) A chasm "intellectually impassable"; and yet
        (*b*) "By an intellectual necessity"
        He crosses it.

IV. The Homers of Modern Materialism
    Buchner, Oken, Haeckel, Huxley.
    "— quandoque bonus dormitat Homerus."

CHAPTER VII.

PROTOPLASM.

Origin of the word.
    The Physiological Cell Theory.
    The several stages which marked the
Application of the word.
    Dujardin, Von Mohl, Cohn, Remak, Max Schultze.
    Prof. Huxley's employment of it to denote
"The Physical Basis of Life:"
    "The one kind of matter which is common to all living beings," and
    Ultimately resolvable into the same chemical constituents.
Ulterior Assumptions:
    By which Protoplasm, From being the "basis"
        Becomes the "Matter of Life."
    That all organisms consist alike of the same "matter of life."
    That this "matter of life" is due to Chemistry alone.
        That all the activities of life,—
            Thought, Conscience, Will,
    Arise solely from,—
        "The arrangement of the molecules of ordinary matter."
MATERIALISM of Mr. Huxley's doctrine.
    In what sense disavowed by him.
    Refuted by Dr. Stirling.
    His admission, that "Most undoubtedly the terms of his propositions are distinctly materialistic."
        E.g., "The thoughts to which I am now giving utterance, and your thoughts regarding them, are but the expression of molecular changes in that matter of life which is the source of our other vital phenomena."
    Mr. Huxley's doctrine, then, is "distinctly materialistic"
    But,—

Is it True?

"I know of no form of negation sufficiently explicit, comprehensive, and emphatic, in which to reply to this question." (*Dr. Elam*)

I. It is in no sense true that Protoplasm "breaks up," as Prof. Huxley says it does.

II. (CO_2), (H_2O), and (NH_3) cannot, by any combination, be brought to represent

$C_{36}H_{26}N_4O_{10}$, which is the equivalent of protein, or protoplasm.

III. It is not true that when carbonic acid, water, and ammonia disappear,

An "equivalent weight of the matter of life" makes its appearance.

IV. In the two processes which Mr. Huxley regards as identical

(*i.e.*, the formation of water and of protoplasm) "There is no resemblance whatever."

V. The proposition that Life is a product of Protoplasm Is demonstrably untrue.

VI. The proposition that life is a property of Protoplasm Is equally untrue.

(Contrast between "aquosity" and "vitality.")

VII. Martinus Scriblerus Redivivus.

The "development" of meat-jacks.

VIII. The identity of Protoplasm, "living or dead,"

Assumed by Mr. Huxley.
Denied by the Germans.
Involves a whole train of Effects without a Cause.

IX. The fulcrum on which Mr. Huxley's Protoplasmic Materialism rests

Is a single inference
From a chemical analogy.
This analogy has two references, and fails in both of them.
The relation of the organic [protoplasm] to the inorganic [water]
Is not an analogy, but an antithesis.
The gulf between Death and Life.

X. The entire Theory
    Summed up in two Propositions.
    "Protoplasm is the clay of the Potter"
    The bricks are the same (says Mr. Huxley)
        Because the clay is the same.
    But—
    Is the clay the same?
    Can it be identified? as Mr. Huxley affirms.
    Examination of the alleged three-fold unity, Faculty, Form, Substance.
    Instead of "identity" there is
        "An infinite diversity."

XI. Protoplasm not convertible
    As alleged by Mr. Huxley.
    Functions, too, are inconvertible, and are
    Independent of mere chemical composition.

XII. As of the Bricks, then, so of the Clay:
    It is not identical
    It is not convertible
    It includes—
        "An Infinitude of various Kinds."

XIII. Mr. Huxley's Variations:
    A complete Revolution of Opinion.

XIV. His "subtle influences"
    Invoked to supersede "Vitality."
    The Bases of Physical Life = (?)
    The Physical Basis of Life
    Cf. "The iron basis of the candle," with
    "The basis of the iron candle"!

XV. His Refutation by Dr. Beale.
    "I doubt if in the whole range of modern science it would be possible to find an assertion more at variance with facts familiar to physiologists."

XVI. His former maintenance of
    "Vitality" and "Inertia."

XVII. Dogmatism of his assertions: Contrasted with Magnitude of his admissions.

Analytical Outline of Contents. xv

XVIII. Dr. Elam's exposure of his Chemistry.
"Professor Huxley's 'Chemistry of Life' has no foundation except that of deliberate and reiterated assertion."

XIX. "Ex ore tuo."
"That such verbal hocus-pocus should be received as science will one day be regarded as evidence of the low state of intelligence in the nineteenth century."

CHAPTER VIII.
THE THREE BEGINNINGS.

Evolution not Eternal.
    The "First Beginnings (Lucretius).
Importance of the Fact:
    There was "a first start"
    There was more than one.

I. 1. MATTER.
    How? Where? Whence? did it Begin
    Its Nature
    Its Properties
    Its Powers
        From what Source acquired?
        "In the Beginning?"
    "The Atoms eternally falling."
    Why "falling?"
    In an eternity "not eternal.
    What Force was that which moved them?
    What Will was that which directed them?

2. Force:
    Operating in a given Order: and
    Controlled by "Definite Laws."
        ORDER: FORCE: LAW:
        How came they to Begin?

3. "Mutual Interaction:
    Of the molecules of the Primitive Nebulosity"
    The sole and exclusive cause of "the whole world; living and not living."

When these assumptions have been granted:
 That the Nebulosity was real
 That it was Primitive
 That its constitutent molecules were not all imaginary
 That the existing world is the result of their interaction
Then, the first question is more urgent than before:
 "In The Beginning:" What was that

4. First Cause:
 Equal, not only to the
 Origination of Matter and of Force, but
 Equal also to the
 Origination of Matter thus constituted, and of Force thus adjusted?

5. Evolution: is thus seen to be the measure of
 Involution.
 Whatever has been evolved in the Effect
 Was previously involved in the Cause.

6. Causa Causarum: What was that?
 In "The First Beginning"?

II. LIFE.

 "Of the causes which have led to the origination of living matter, it may be said that we know absolutely nothing." (Huxley)
 But, however inscrutable the mode,
 There is no question, nor any room for question
  As to the Fact:
 "Living matter" was "once originated."
 Life had a BEGINNING.

 Still more inscrutable is the Mystery which shrouds
 The First Emergence of

III THE SELF-CONSCIOUS MIND.
 Mr. J. S. Mill on the Existence of Mind.
 Huxley, Tyndall, and Spencer, on "States of Consciousness."
 "Consciousness," says Prof. Huxley, is "unaccountable."
 "No one can prove that mind and life are in any way related to chemistry and mechanics."

Analytical Outline of Contents. xvii

Consciousness and Physics are incommensurable.
"Thought BEGAN to be." How?
"Intelligence, self-conscious, *emerged.*"
                WHENCE?

CHAPTER IX.

THE THREE BARRIERS.

Mr. Darwin on
    The adaptation of organs,
    The transmutation of animals,
    The Origin of Instinct,
    The ant, and the honey-bee.
His Theory of Neuters :
    Fertile parents transmit,
        through fertile progeny,
    A tendency to produce sterility,
        incapable of further production.
His oversight of
    The evidence of Design.
    His Remarkable Omissions.
His ingenious substitution of
    The "conceivable" for the actual.
His habitual avoidance of
    The profounder marvels of Nature, and
    Their only true solvent—
        The ordination of God.
The Three Barriers of
    Comparative Anatomy.
I. THE BACKBONE :
    The basis of Strength.
    An impassable Barrier
    Until it can be shewn
        How a butterfly could become a bird,
        Or a snail, a serpent,
        Or a star-fish acquire the skeleton of
            a salmon or a shark.

xviii *Analytical Outline of Contents.*

II. THE BREAST:
>The type of Tenderness
>Until it can be shewn
>>How an animal that never was suckled stumbled on the capacity of *giving* what it never got.

III. THE BRAIN:
>The measure of Capacity.
>. The Human Brain is Pleno-cerebral:
>All other Brains are Manco-cerebral.
>To *all* Men the pleno-cerebral type is *common:*
>To *Man*, as such, it is PECULIAR.
>The lowest Human Brain has the latent franchise of
>>Progressive Reason:
>All other Brains have the rigid circumscription of
>>Unprogressive Instinct.
>No brute is susceptible of Human Culture;
>No human infant is not so.
>Between these two the Difference is Immeasurable

CHAPTER X.
ATOMS.

"The Atoms are the First Beginnings."
What, then, are these Atoms?
>"Ultimate homogeneous units;"
>Lange. Mr. Herbert Spencer.
>"One ultimate form of Matter."

Dr. Tyndall's rejection of
>Mr. Spencer's dictum.

Heterogeneity of the Atoms.
>Chemical Phenomena
>>Not to be deduced from
>Mechanical conditions.

Their grouping: Their varieties:
>In shape; In kind.

Their Motions, Forces, Affinities:
>Inadequate to the problem proposed.

Analytical Outline of Contents. xix

The "Atoms" are
    NOT the Beginning.
    They have "all the characteristics of
MANUFACTURED ARTICLES."
    Sir John Herschel.
"No Theory of Evolution can be formed to account for them."
    Professor Maxwell. Professor Pritchard.
Sir William Thomson :—
        "The assumption of atoms can explain no property of body which has not previously been attributed to the atoms themselves."

CHAPTER XI.

APES.

Professor Tyndall's postulate :—
    That human ancestors were not human.
Mr. Darwin's :—
    "A series of forms graduating insensibly
    From some ape-like creature
    To man as he now exists." But
    (i.) The series is not a series.
        It has no continuity, and no concatenation.
    (ii.) It does not "graduate insensibly."
        It exhibits "breaks": "wide, sharp, and defined."
        These breaks "incessantly occur in all parts of the series."
    (iii.) The "ape-like creature" is wholly hypothetical.
        It is absolutely non-existent.
        There is no evidence that it ever was otherwise.
Professor Huxley's
    Cautious and conditional generalizations
    Adverse to Mr. Darwin's theory.

Professor Haeckel's
    "Rogues in buckram."
    Destitute of any single living representative.
    Destitute of fossil evidence of their former existence.
The *Chordonia*
    "*Developed* THEMSELVES"!
The admissions of its advocates, are
    Fatal to The Theory.

CHAPTER XII.
MEN.

Prof. Huxley's *dicta* on
    "The question of questions for mankind."
Contrast between Men and Apes:
    As to cerebral *structure*.
    As to cerebral *weight*.
    As to "the great gulf in intellectual power
        between lowest man and highest ape."
    As to "the structural differences
        which separate Man from the Gorilla."
No intermediate Link
    bridges over the gap between *Homo* and *Troglodytes*."
Paradoxes:
    "Quâ-quâ-versal propositions."
"The UNMEASURABLE and *practically infinite* divergence
    Of the Human from the Simian Stirps."
    Its "Primary Cause."
Psychical Distinctions.
Structural Distinctions.
Mr. Darwin's Testimony to
    "The great break in the organic chain
    Between man and his nearest allies, which
    Cannot be bridged over
    By any extinct or living species."
Prof. Mivart's Refutation of this theory.
    Man, the apes, and the half-apes
    Cannot be arranged in a single ascending series.

Analytical Outline of Contents. xxi

The Lines of Affinity existing between different Primates
    Construct a network; but not a ladder.
The Survival of the Fittest.
    But the fittest (according to the Theory)
    Have not survived.
    The half-apes are with us to this day:
    The half-men are nowhere.
Mr. Wallace's Demonstration
    That the Origin of Man is to be found only in
    An Act of Special Creation.
Mr. Mivart's Conclusion:
    That Mr. Darwin "has UTTERLY FAILED
        In the only part of his work which
        is really important."

CHAPTER XIII.

ANIMA MUNDI.

"A Soul in all things."
    The Inorganic World.
    Phenomena of Crystallization.
    Prof. Tyndall's Fallacy;
    Pyramid builders: Architect: Controlling Power.
Prof. Tyndall's belief that
    "The formation of a plant or an animal
    Is a *purely mechanical* problem."
Prof. Huxley's assertion that
    "A mass of living protoplasm
    Is simply a molecular machine."
His resort to "subtle influences,"
    i.e., to Vital Force.
His assertion that
    "A particle of jelly" *guides* forces.
    Refuted by Dr. Beale.
Two Points involved in these assertions:—
    I. The introduction of Life;
    II. The manifestations of Mind.

xxii *Analytical Outline of Contents.*

I. VITAL ACTION : In contrast with physico-chemical action
   Is peculiar to living beings.
   Haeckel's Testimony :—
       "The phenomena which living things present have no parallel in the mineral world."
   Du Bois Reymond's :—
       "It is futile to attempt by chemistry to bridge the chasm between the living and the not-living."
   No machine can *grow.*
   No machine can *produce machines like itself.*

II. MIND. 1. "Horology" : and the "watch-force" :
   A combination of many forces, and
   Their adjustment to a particular PURPOSE.
   Its seat is in
       The Intelligence which conceived that combination; and in
       The Will which gave it effect.
   This evidence of Design is shewn in Universal Nature.
   2. The Shell of the Barnacle.
   3. The Electric Ray.
       "It is impossible to conceive by what steps these wondrous organs have been produced." (Mr. Darwin.)
   4. The new-born Kangaroo.
       "Irrefragable evidence of Creative foresight." (Prof. Owen.)
   5. The Eye : "*With all its* INIMITABLE *contrivances.*" (Mr. Darwin) (Prof. Pritchard.)
Nature is full of Plan.
   Yet she plans not.
   Where Science assumes a Use,
       Religion affirms an Author.
   The Question, *For what?*
   Involves the further question, *From whom?*

Analytical Outline of Contents.

Mr. Ruskin, on The Great First Cause
    "Personal": and "A Supporting Spirit in all
        things."
    The Formative Cause.
    The Living Power.
    What is it? and *Whence?*
    "There is no answer."
Ascensive Life.
Language : Peculiar to Man :
    "Thinker of God's thoughts after Him."
What is the Origin of Mind?
The genesis of THOUGHT.
    "Thaumaturgic." (Carlyle.)
    "No mere function of The Brain."
    "A World by itself."
VOLITION. Whence?
    A machine not mechanical.
    "An automaton endowed with free will."
CONSCIOUSNESS.
    "A rock on which Materialism must inevitably
        split." (Tyndall.)
    Perfectly "unaccountable." (Huxley.)
    "Brain-waves." (Ruskin.)
SENSE OF RESPONSIBILITY.
    "Duty! . . . WHENCE THY ORIGINAL?
    (Kant.)
THE MAJESTIC SPECTACLE OF THE UNIVERSE
    Is a spectacle for the eye of Reason.
    Natural Agents working for ends *which they them-*
        selves cannot perceive.
    But "Every house is builded by some man";
        And
"HE THAT BUILT ALL THINGS, IS GOD."

CHAPTER I.

THE RIGHT OF SEARCH.

"GOD created man"? No such thing! The monads developed him. "The heavens declare the glory of God"? Far from it: "they declare only the glory of the astronomer!" "We have now no need of the hypothesis of God."

These utterances, and such as these, startling alike to reverence and to faith, are the merest common places of modern agnosticism. Instead of being, as once they were regarded, the *terminus ad quem*, the ultimate goal, to which unbelief was tending, they have long since been left behind as a mere *terminus à quo*, a temporary station for a new point of departure. The scepticism which doubted has given place to the dogmatism which denies. "Honest doubt" has been supplanted by the clamour of a positive self-assertion. A positivism of which Comte knew nothing has usurped the authority, while renouncing the functions, of scientific enquiry.

In a word, Agnosticism is no more, and Gnosticism reigns in its stead.

Agnosticism made candid confession of its ignorance. Gnosticism parades its pretensions to knowledge. The former did not know: the latter is quite sure. The Divine existence is now declared to be not only unnecessary; it is absolutely unreal. God has no existence, even hypothetically, except as the creature of the human imagination. The hand may well tremble that writes it, and the ears may tingle that hear, yet it has been both written and said—in modes that demand more attention than they have hitherto received—There is no God! except such as man has made. "The dim and shadowy outlines of the superhuman deity fade slowly away from before us; and as the mist of his presence floats aside, we perceive with greater and greater clearness the shape of a yet grander and nobler figure—of Him who made all gods and shall unmake them."[1]

Who then is He, this "grander and nobler figure," this great and only potentate "who made all gods and shall unmake them"? this "human" who dethrones "the superhuman deity"? It is man himself. "From the dim

[1] Professor Clifford: "The Ethics of Religion," in *The Fortnightly Review*, vol. xxii. New series, p. 52.

dawn of history, and from the inmost depth of every soul, the face of our father Man looks out upon us with the fire of eternal youth in his eyes, and says, 'Before Jehovah was, I am!'"[1]

And yet, this "Man our father," was once an Ape: and, before that, "a jelly-bag." That jelly-bag (which "made all gods and shall unmake them") sucking in water and sticking to a stone, has advanced to its present august condition by "a principle of development" and "a process of evolution." It is true indeed that the principle is one which nobody has ever proved, and the process is one which nobody has ever witnessed; but woe to the unlucky wight who dares to doubt their validity, or who fails to recognise in "Mr. Charles Darwin, the Abraham of scientific men."[2]

"Most of you," says Professor Tyndall, "have been forced to listen to the outcries and denunciations which rang discordant through the land for some years after the publication of Mr. Darwin's 'Origin of Species.' Well, the world—even the clerical word—has for the

---

[1] Professor Clifford: "The Ethics of Religion," in *The Fortnightly Review*, vol. xxii. New series, p. 52. *Vide infra:* Appendix, Note A.

[2] Prof. Tyndall: "Science and Man," in *The Fortnightly Review*, vol. xxii. New series, p. 615.

most part settled down in the belief that Mr. Darwin's book simply reflects the truth of Nature: that we who are now 'foremost in the files of time' have come to the front through almost endless stages of promotion from lower to higher forms of life."[1]

"The most part": but what of the rest, the remaining part? Let it stand in awe. If it cannot be convinced it can be denounced. And it is denounced accordingly. It is more base and stupid than—"even the clerical world." He who belongs to it is *ipso facto* stigmatized as ignorant and incompetent.[2] He is "unstable and weak,"[3] "a brawler and a clown."[4]

---

[1] Prof. Tyndall: "Science and Man," in *The Fortnightly Review*, vol. xxii. New series, p. 611.

[2] The great and venerated name of Von Baer is associated by Haeckel with the idea of "harmless senile garrulity." Adolf Bastian is a "Privy Councillor of Confusion"; Du Bois-Raymond is a "rhetorical phrase-spinner," if not a Professor of Voluntary Ignorance; while Carl Semper is a—a person regardless of truth, expressed in a brief word not usually heard among gentlemen. "Haeckel," says Dr. Elam, "has probably never heard of the insignificant names of Owen, Mivart, and Agassiz, or they would doubtless have been remembered in the catalogue of wretched smatterers who have come under his signal disapproval."

[3] Prof. Tyndall's "Address delivered at Belfast." Longmans, 1874, p. 63.

[4] *Fortnightly Review*, vol. xxii. p. 614.

But "methinks the lady doth protest too much." Were these denunciations more dispassionate they might seem more disinterested. As it is, they are too strenuous to be forcible; too loud to be effective. Nor is this the worst They have another fault more fatal still. They are altogether irrelevant. They do not hit, they merely miss, the mark. They are beside the question. For the question is as to the nature and character of the new doctrine. And with that question the merits or demerits of advocates and assailants are not concerned. "Materialistic Atheism," we are told, "is in the air." So be it: but then this same materialistic atheism is either true or it is not. If it is not true, let that be shown, and it will fall without assailants. If it is true, let that be shown, and it will then have no need of advocates. No one thinks it necessary to take the field in defence of the properties of conic sections; and the foundations of the venerable *pons asinorum* remain unmoved and unimpaired from age to age. Why then, in propounding that very open secret, their latest discovery, should the demigods of the scientific Olympus forsake their philosophic calm for the irritating gusts of irascible acerbity?

Tantæne animis cœlestibus iræ?

They make their boast of truth. They proclaim aloud their contempt of consequences. The boast would have been more becoming if it had been less exclusive. Those who make it will have a better claim to be heard when they have learned, with the modesty of science, to moderate the pretensions by which they arrogate to themselves a monopoly of the virtue which they say is theirs. When they tell us that "Mr. Charles Darwin, the Abraham of scientific men," is "a scholar as obedient to the command of truth as was the patriarch to the command of God,"[1] we are under no necessity, as we certainly have no inclination, to dispute the accuracy of the assertion. But when to this it is added that to reject Mr. Darwin's hypothesis, and those of his coadjutors and commentators, is "to purchase intellectual peace at the price of intellectual death,"[2] we ask for the evidence in support of this assertion. That evidence has yet to be produced. Is it producible? It is at all events not forthcoming. Until the truth of these hypotheses has been established it is not possible, in the name of truth, to demand our acceptance of them. And until then, as always, our position in relation to

[1] *Fortnightly Review*, vol. xxii. p. 615.
[2] "The Belfast Address," *ut sup.*, p. 63.

them must be determined, as it is now determined, by that paramount consideration, our reverence for truth.

The necessity of meeting this conviction is not unfelt by those to whom it is opposed ; and their perception of its force is shown by the remarkable admission contained in their reply. It is the ideal Lucretian himself who is the speaker :—

"It is not to the point to say that the views of Lucretius and Bruno, of Darwin and Spencer may be wrong. Here I should agree with you, deeming it indeed certain that these views will undergo modification. But the point is, that whether right or wrong, we ask the freedom to discuss them."[1] "As regards these questions science claims unrestricted right of search."[2]

Agreed. We desire nothing better. The case must be argued before it is decided. And it may not be prejudged. What is certain is, "that the views of Lucretius and Bruno, of Darwin and Spencer may be wrong": "certain that these views will undergo modification." Certain therefore that "the world,—even the clerical world,"—in accepting these wrong views, "has for the most part" gone wrong too, and,

[1] "The Belfast Address," *ut sup.*; p. 64.
[2] *Ibid.*, p. 63.

sooner or later, not without harm and loss, will have to return from the error of its ways.

Meantime, the inquiry to which we are challenged, though not without complex relations, is in itself very simple. It is not to be influenced by opinion. It is not to be biassed by prejudice. It is not to be decided by authority. It is directed to the investigation of facts. It must be guided, not by great names, but by great principles. It must be kept distinct from other, though collateral, inquiries; and it must be patiently pursued to no uncertain issue. This Materialistic Atheism, propounded in the name of Science: Is it true? Is it demonstrable? Is it Scientific?

CHAPTER II.

EVOLUTION.

IT stumbles at starting. Of Evolution as alleged, there are several varieties; and the theory is at fault among them. A choice must be made, and the choice is not easy. Natural Selection, if it were not merely the nominal designation of an unreal entity, might here render important service; but as it is, is useless. And to spontaneous selection the choice is encumbered with difficulties. Of these difficulties it is not the least that, by the theory, spontaneous selection is impossible: spontaneity is non-existent, save in imagination. Since this little difficulty is not (by the theory) to be surmounted, it must be evaded; and when it has been evaded the labour of selection begins.

The varieties from which the selection must be made may be classed in three main divisions; or, in other words, notwithstanding the protests of those Darwinians who deny the existence of species, they may all be referred to three species: the theistic, the atheistic, and the agnostic.

Evolutionists of the first class admit, while those of the second deny, the existence of a Divine Creator. By those of the third class, that existence, while not by any means admitted, is yet not explicitly denied. It is simply ignored. They "have no need of the hypothesis of God." Foremost among the leaders of this latter class are Mr. Spencer and Professors Huxley, Tyndall, and Bain. Less cautious or more candid are Carl Vogt, Ernst Haeckel, and Buchner, as representatives of atheistic development; while the theistic, its antithesis, is vindicated by names of no less note than those of Sir John Herschel, Sir William Thomson, Professors Owen, Dawson, Gray, Dr. Carpenter, and, at least in his earlier writings, Mr. Charles Darwin himself.

The existence of these varieties is a fact at once significant and instructive. Our present concern, however, is not with these, except so far as they serve to illustrate or demonstrate the nature of the base which they have in common. That doctrine of Development which they all affirm: what is it? What are its pretensions? Where are its proofs?

Evolution. 35

Let "the Abraham of scientific men" speak first.

"It is interesting," he says,[1] "to contemplate an entangled bank, clothed with many plants of many kinds, with birds singing on the bushes, with various insects flitting about, and with worms crawling through the damp earth, and to reflect that these elaborately-constructed forms so different from each other, and dependent on each other in so complex a manner, have all been produced by laws acting around us." "There is grandeur in this view of life, with its several powers, having been originally breathed into a few forms or into one."

The grandeur, however, is questionable. It may be nothing more than a figment of the imagination, a mere matter of taste, or of opinion; but even if it were matter of fact, it is not a matter with which we have any concern. Our enquiry as to "this view of life" is not, Can it be made to look grand? but, Can it be shown to be true?

At present, this has not been shown. Even Mr. Darwin himself does not profess to "know," he merely "believes," the truth of the doctrine he propounds. "I believe," these are his words,

[1] "Origin of Species." First Edition (Murray : 1859), chap. xiv. pp. 489, 490.

"that animals have descended from at most only four or five progenitors, and plants from an equal or lesser number. Analogy would lead me one step further, namely, to the belief that all animals and plants have descended from some one prototype. But analogy may be a deceitful guide. Nevertheless all living things have much in common, . . . Therefore I should infer from analogy that probably all the organic beings which have ever lived on this earth have descended from some one primordial form into which life was first breathed."[1]

But this "belief," which Mr. Darwin thinks "probable," this "inference" derived from "analogy," has never been verified. How could it be verified, when its most ardent apostles assure us that it may, after all, "be wrong," and will "certainly" have to "undergo modification?"[2] But even if it had been verified it is not "materialism," it is not "atheism," it is not "agnosticism." It is the very reverse of all these, for it is a manifesto of absolute "theism."

"In my book on the 'Genesis of Species,'" says Professor St. George Mivart,[3] "I had in

---

[1] "Origin of Species." First Edition, chap. xiv. p. 484.
[2] Prof. Tyndall, *ut sup.*, p. 7.
[3] "Lessons from Nature." Murray, 1876, p. 429.

view two main objects. My first was to show that the Darwinian theory is untenable, and that 'Natural Selection' is not *the* origin of species. My second was to demonstrate that nothing even in Mr. Darwin's theory (as put forth before the publication of his 'Descent of Man,') and, *à fortiori*, nothing in Evolution generally, was necessarily antagonistic to Christianity."

Reserving for further examination the first of these propositions, "that the Darwinian theory is untenable," it may be observed as to the second, that of the theistic doctrine of Evolution there are theoretically three main varieties: (1) That which limits the supernatural action in the origination of species to the creation of primordial cells. (2) That which, while maintaining the intervention of direct or special creation, regards the origination of species as being for the most part effected indirectly, *i.e.*, through the agency of natural causes. (3) That which regards God as immanent in natural law, and recognises in all phenomena the result of present Divine action.

In his earlier writings, the theism of Mr. Darwin is most explicit. Thus, for example, when speaking of certain birds found in Tierra del Fuego, he says, "when finding, as in this case, any animal which seems to play so insignificant a part in the great scheme of nature, one is apt to wonder why a distinct species should have been created; but it should always be recollected that in some other country perhaps it is an essential member of society, or at some former period may have been so."[1] And again: In his description of the Passage of Cordillera, he says, "I was very much struck with the marked difference between the vegetation of these eastern valleys and that of the opposite side: yet the climate, as well as the kind of soil, is nearly identical, and the difference of longitude very trifling. The same remark holds good with the quadrupeds, and in a lesser degree with the birds and insects." "This fact," he adds, "is in perfect accordance with the geological history of the Andes; for these mountains have existed as a great barrier since a period so remote that whole races of animals must subsequently have perished from the face of the earth. Therefore, unless we suppose the same species to have been created in two different countries, we ought not to expect any closer similarity between the organic beings on opposite sides of the Andes, than

---

[1] "Narrative of the Surveying Voyages of H.M.'s Ships *Adventure* and *Beagle*." London, 1839. Vol. iii.

Evolution. 39

on shores separated by a broad strait of the sea."[1]

And to take but one other instance: In concluding his review of the causes of extinction of certain animals in Patagonia, he says,—" We see that whole series of animals, which have been created with peculiar kinds of organization, are confined to certain areas; and we can hardly suppose these structures are only adaptations to peculiarities of climate or country; for otherwise, animals belonging to a distinct type, and introduced by man, would not succeed so admirably even to the extermination of the aborigines. On such grounds it does not seem a necessary conclusion, that the extinction of species, more than their creation, should exclusively depend on the nature (altered by physical changes) of their country."[2] In these passages we have not only the assertion of species as an established distinction in animal life, we have also the further assertion that these "distinct species," "with peculiar kinds of organization," are to be attributed to "Creation" as their cause, and not "to peculiarities of climate or country."

[1] " Narrative of the Surveying Voyages of H.M.'s Ships *Adventure* and *Beagle*." London, 1839. Vol. iii. pp. 399, 400.
[2] *Ibid.*, p. 212.

But in his later works, the theism thus articulately pronounced is conspicuous chiefly by its absence. At the same time it is not expressly excluded. And on this account the agnostic and atheistic leaders take him roundly to task, notwithstanding his Abrahamic dignity. Thus, for instance, Professor Tyndall :—

"Diminishing gradually the number of progenitors, Mr. Darwin comes at length to one 'primordial form;' but he does not say, as far as I remember, how he supposes this form to have been introduced. He quotes with satisfaction the words of a celebrated author and divine, who had 'gradually learnt to see that it is just as noble a conception of the Deity to believe He created a few original forms, capable of self-development into other and needful forms, as to believe that He required a fresh act of creation to supply the voids caused by the action of His laws.' What Mr. Darwin thinks of this view of the introduction of life I do not know. But the anthropomorphism, which it seemed his object to set aside, is as firmly associated with the creation of a few forms as with the creation of a multitude. We need clearness and thoroughness here. Two courses and two only are possible. Either let us open our doors freely to the conception of creative acts, or, abandoning them, let us radically change our notions of Matter."[1]

Professor Tyndall, as is well known, adopts the latter of these alternatives, and discerns in Matter "the promise and potency of all terrestrial life."[2] To do this, however, is, as he himself declares, to "abandon," once for all, "the conception of creative acts."

Has Mr. Darwin abandoned that conception? If he has not, then he lacks "clearness and thoroughness" — "father of scientific men" though he be. So, at least, says Professor Tyndall, and Professor Huxley goes still further.

Mr. Huxley's utterances on this subject possess a special interest from the eulogy pronounced on him as the accredited "expounder" of the Darwinian doctrine. Thus, at Belfast, when introducing his summary of "The Origin of Species," Professor Tyndall said,—

"The book was by no means an easy one; and probably not one in every score of those who then attacked it had read its pages through, or were competent to grasp its significance if they had. I do not say this merely to discredit them; for there were in those days some really

[1] "Address delivered before the British Association at Belfast." Longmans, 1874, pp. 53, 54.
[2] *Ibid.*, p. 55.

eminent scientific men, entirely raised above the heat of popular prejudice, willing to accept any conclusion that science had to offer, provided it was duly backed by fact and argument, and who entirely mistook Mr. Darwin's views. In fact, the work needed an expounder; and it found one in Mr. Huxley. I know nothing more admirable in the way of scientific exposition than those early articles of his on the origin of species. He swept the curve of discussion through the really significant points of the subject, enriched his exposition with profound original remarks and reflections, often summing up in a single pithy sentence an argument which a less compact mind would have spread over pages."[1]

Now the pithy sentence with which we are here concerned is this:—

"The improver of natural knowledge absolutely refuses to acknowledge authority as such. For him, scepticism is the highest of duties, blind faith the one unpardonable sin. The man of science has learned to believe in justification, not by faith, but by verification."[2]

And with this Professor Tyndall agrees: "Without verification a theoretic conception is

[1] "Address," *ut sup.*, p. 38.
[2] "Lay Sermons." Macmillan, 1871, p. 18.

a mere figment of the intellect." Torricelli, Pascal, and Newton were distinguished by their "welding of rigid logic to verifying fact." "If scientific men were not accustomed to demand verification . . . their science, instead of being, as it is, a fortress of adamant, would be a house of clay." "Newton's action in this matter is the normal action of the scientific mind."[1] "There is no genius so gifted as not to need control and verification."[2]

What then becomes of "the Abraham of scientific men"? In the "Origin of Species" Mr. Darwin tells us repeatedly,[3] that it would be "fatal" to his theory if it should be found that there were characters or structures which could not be accounted for by "numerous, successive, slight modifications"; and tnis candid admission is supplemented in the "Descent of Man,"[4] by another equally candid :—

[1] "Fragments of Science." Longmans, 1871, pp. 59, 62.
[2] *Ibid.*, p. 111.
[3] See especially, (First Edition,) p. 189, where, after attempting to explain the origin of the eye, he says, " If it could be demonstrated that any complex organ existed, which could not possibly have been formed by numerous, successive, slight modifications, my theory would absolutely break down.
[4] Murray, 1871, vol. ii. p. 387.

"No doubt man, as well as every other animal, presents structures which, as far as we can judge with our little knowledge, are not now of any service to him, nor have been so during any former period of his existence, either in relation to his general condition of life, or of one sex to the other. Such structures cannot be accounted for by any form of selection, or by the inherited effects of the use and disuse of parts."

Here, then, we have the fullest recognition of the validity of objections which are absolutely fatal to his whole doctrine. But with this recognition, what becomes of "verification"?

Mr. Darwin's doctrine, however, constitutes a very small part of that "theoretic conception" which, under the name of Evolution, is now declared by Professor Huxley to be no longer "a matter of speculation and argument," but on the contrary, has "become a matter of fact and history." "The history of Evolution," he adds, "as a matter of fact, is now distinctly traceable. We know it has happened, and what remains is the subordinate question of how it happened."[1]

It is to be observed, however, that the "Evo-

[1] "Address at Buffalo," August 25th. Reported in *The Times* of Sept. 14, 1876.

Evolution. 45

lution" of which Mr. Huxley makes this affirmation, is something very different from the indefinite nondescript which in popular writings is often designated by the same term. Not unfrequently "evolution" means simply progress or advancement. It is even used when nothing more than growth is intended. It is employed as if it were identical with "natural selection," or "transmutation," or any other mode of "development." But with Mr. Huxley, evolution is something more than the emergence of the chick from the egg, or the oak from the acorn, or the frog from the tadpole. It is not a mere increase of bulk, nor is it restricted to any particular process, nor has it any special aim. It is a change from simplicity to complexity; from incoherence and indefiniteness to their opposites.

Thus, for instance, the nebular hypothesis supposes the evolution of the planetary bodies from incoherent atoms, which come not merely into mutual relation, but which also in that process become grouped together in such a way that the nascent mass becomes complex, consists of parts. Again: the homogeneous protoplasm in which all organized beings commence, shows, when under favourable conditions, first the elements of tissues. These elements are afterwards grouped into tissues, and the tissues are associated into organs. The "indifferent" matter is differentiated in various degrees, and the animal and vegetable series show many grades of difference.

Thus the Protamœba never reaches to the formation of tissues; the Hydra has tissues, but few organs; and, ascending in the series, the sharks, complex as is their organization, exhibit a less thorough differentiation of their hard parts, which are chiefly cartilaginous, than do mammals, in which cartilage is subordinate to bone. But the evolution of the more complex from the more simple organisms does not necessarily form a linear series; probably it never does so. Nor does evolution imply change of matter as well as of the relations of its parts; fresh matter is not essential to it, since the phenomena which it includes are, as matter of fact, rearrangements of that which was already existing.

Such are the principal facts regarding evolution; and from these it is evident that the phenomena themselves are absolutely independent of any and of every theory as to their cause. Thus understood and thus limited, Evolution,—*i.e.*, the phenomenal sequence, not the ideal hypothesis—is a law the operation

Evolution. 47

of which is traceable throughout every department of nature.

Mr. Herbert Spencer's definition of it is equally clear and concise: "Evolution is a change from an indefinite, incoherent homogeneity, to a definite, coherent heterogeneity; through continuous differentiations and integrations."[1]

Its absolute universality of operation he thus expresses: "Whether it be in the development of the Earth, in the development of Life upon its surface, in the development of Society, of Government, of Manufactures, of Commerce, of Language, of Literature, Science, Art, this same advance from the simple to the complex, through successive differentiations, holds uniformly. From the earliest traceable cosmical changes down to the latest results of civilization, we shall find that the transformation of the homogeneous into the heterogeneous, is that in which Evolution essentially consists."[2]

In this last sentence we have not merely the "transformation" "in which evolution essentially consists;" we have also the assump-

---

[1] "First Principles." Williams & Norgate, 1862, p. 216. A subsequent definition is given below. See Appendix, Note B.
[2] *Ibid.*, pp. 148, 149.

tion that "the latest results of civilization" have been evolved, in the way of necessary and inevitable consequence, from "the earliest traceable cosmical changes." Human life, with all its inexhaustible possibilities, has been evolved from life infra-human. The life of the lower animals, like that of plants, was in the first instance evolved from non-living matter; as that matter itself was evolved from "cosmic vapour."

Professor Tyndall, as we have seen, tells us that "the world—even the clerical world—has for the most part settled down in the belief that Mr. Darwin's book simply reflects the truth of Nature: that we who are now 'foremost in the files of time' have come to the front through almost endless stages of promotion from lower to higher forms of life."[1] And again:—

"It is now generally admitted that the man of to-day is the child and product of incalculable antecedent time. His physical and intellectual textures have been woven for him during his passage through phases of history and forms of existence which lead the mind back to an abysmal past."[2] "If to any one of us were given the privilege of looking back through the

[1] "Science and Man." *Fortnightly Review*, vol. xxii. p. 611.
[2] *Ibid.*, p. 594.

Evolution.

æons across which life has crept towards its present outcome, his vision would ultimately reach a point when the progenitors of this assembly could not be called human."[1] "No one indeed doubts now that all the higher types of life with which the earth teems have been developed by the patient process of evolution from lower organisms, and in logical consistency we are bound to trace back the series to the simplest forms of protoplasm, which the microscope reveals to us as living units. But all this is but the outcome of life from life, and leaves us without an approach to a solution of the mighty question of the origin of life. There was a time when the earth was a red-hot melted globe, on which no life could exist. In course of ages its surface cooled; but, to quote the words of one of our greatest *savans*, 'when it first became fit for life there was no living thing upon it.' How then are we to conceive the origination of organized creatures?"[2]

Professor Huxley, propounding to the British Association[3] the tenets of what he called his

---

[1] "Science and Man." *Fortnightly Review*, vol. xxii. p. 611.
[2] "The Germ Theory and Spontaneous Generation." *Contemporary Review*, vol. xxix. pp. 901, 902.
[3] In the Presidential Address for 1870.

"philosophic faith" on this subject, has answered this question with his characteristic clearness of enunciation :—

"If it were given me to look beyond the abyss of geologically recorded time to the still more remote period when the earth was passing through physical and chemical conditions, which it can no more see again than a man can recall his infancy, I should expect to be a witness of the evolution of living protoplasm from not living matter."[1]

To the same effect, and not less articulately, Professor Tyndall :—

"The problem before us is, at all events, capable of definite statement. We have on the one hand strong grounds for concluding that the earth was once a molten mass. We now find it not only swathed by an atmosphere and covered by a sea, but also crowded with living things. The question is, how were they introduced? . . . The conclusion of science, which recognises unbroken causal connection between the past and the present, would undoubtedly be that the molten earth contained within it the elements of life, which grouped themselves into their present forms as the planet cooled. The difficulty and reluctance encoun-

[1] "Critiques and Addresses." Macmillan, 1873, p. 239.

tered by this conception, arise solely from the fact that the theologic conception obtained a prior footing in the human mind. Did the latter depend upon reasoning alone, it could not hold its ground for an hour against its rival. . . . Were not man's origin implicated, we should accept without a murmur the derivation of animal and vegetable life from what we call inorganic nature. The conclusion of pure intellect points this way and no other." [1]

In other words—and to sum up all that has been said in one short but authoritative sentence—" The doctrine of Evolution derives man in his totality from the interaction of organism and - environment through countless ages past." [2]

And this it does, whatever may become of Darwinism. On this head, as well as on the illimitable sphere of its operation, we have the final conclusion of Professor Huxley :—

"But even leaving Mr. Darwin's views aside, the whole analogy of natural operations furnishes so complete and crushing an argument against the intervention of any but what are termed

[1] "Materialism and its Opponents." *Fortnightly Review*, vol. xviii. pp. 596, 597.
[2] Prof. Tyndall's "Belfast Address," p. 59.

secondary causes in the production of all the phenomena of the universe ; that in view of the intimate relations between Man and the rest of the living world; and between the forces exerted by the latter and all other forces, I can see no excuse for doubting that all are co-ordinated terms o Nature's great progression, from the formless to the formed,—from the inorganic to the organic,—from blind force to conscious intellect and will."[1]

[1] "Evidence as to Man's Place in Nature." Williams and Norgate, 1803, p. 108.

CHAPTER III.

"*A PUERILE HYPOTHESIS.*"

THIS, then, is Evolution: "baldest of all the philosophies which have sprung up in our world." The evolution which solves the problem of human origin by the assumption that human nature exists potentially in mere inorganic matter; and the assertion that man, with all his powers, and all their products, is the necessary result, by spontaneous derivation, of the interaction of incandescent molecules.

But is this evolution scientific? Is it demonstrable? Is it true? Before this question its assumptions cannot save it, however large; its assertions cannot prove it, however loud. The question lies deeper. Has it received the necessary "verification?" The "verification" without which, however ingenious as a theoretic conception, it must ever remain "a mere figment of the intellect?"[1]

[1] Prof. Tyndall's "Fragments of Science," p. 469.

To this question the answer is both unambiguous and conclusive. To present it the more clearly, let us take separately the two points involved. First, what is the evidence for the succession of life from lower to higher forms? And second, what is the evidence as to the existence of any instance of the conversion or transmutation of one species into another?

Let Professor Huxley answer. For we shall find no witness more competent than he; none whose authority in all matters of natural history and palæontology is more indisputable; none more illustrious in his championship of Evolution in general, or of Mr. Darwin's views in particular. "There is but one hypothesis," he tells us, "as to the origin of species of animals in general which has any scientific existence—that propounded by Mr. Darwin."[1] Testimony from that quarter, therefore, cannot fail to have a special force. And on the first part of the question Professor Huxley writes thus:—

"What, then, does an impartial survey of the positively ascertained truths of palæontology testify in relation to the common doctrines of progressive modification, which suppose that modification to have taken place by a necessary progress from more to less embryonic forms, or from

---

[1] "Man's Place in Nature," p. 106.

more to less generalized types, within the limits of the period represented by the fossiliferous rocks?

"It negatives those doctrines, for it either shows us no evidence of such modification, or demonstrates such modification as has occurred to have been very slight; and as to the nature of that modification, it yields no evidence whatsoever that the earlier members of any long-continued group were more generalized in structure than the later ones. . . .

"Contrariwise, any admissible hypothesis of progressive modification must be compatible with persistence without progression through indefinite periods."[1]

In other words, the "hypothesis" requires some proof of "progressive modification," but it receives none. What it does receive is disproof only. To its demand for "progression," "the fossiliferous rocks" reply by exhibiting only "persistence without progression;" and that, "through indefinite periods." To its assumption of "almost endless stages of promotion from lower to higher forms of life,"[2] Palæontology responds by demonstrating that of these "stages" there is "no evidence," and of this "promotion" there is "no evidence whatsoever."

Nor does Professor Huxley stop here. Dealing with the supposition that such a hypothesis as that of progressive modification should "even-

[1] "On Persistent Types of Life:" in "Lay Sermons," p. 225.
[2] Prof. Tyndall's "Science and Man."

tually be proved to be true," he makes the important statement that the only way in which it can be demonstrated will be "by observation and experiment upon the existing forms of life."[1] But demonstration of this kind is non-existent. Abundantly and incessantly as it has been attempted, it has never yet been achieved. Tried by this test of "observation and experiment upon the existing forms of life," neither Organic Evolution in general nor Mr. Darwin's "Origin of Species" in particular, has any actual place *in rerum naturâ.*

On the second part of the question—that of the transmutation of species—Mr. Huxley writes :—

"After much consideration, and with assuredly no bias against Mr. Darwin's views, it is our clear conviction that, as the evidence stands, it is not absolutely proven that a group of animals, having all the characters exhibited by species in nature, has ever been originated by selection, whether artificial or natural."[2] And again :—

"Our acceptance of the Darwinian hypothesis must be provisional so long as one link in the chain of evidence is wanting ; and so long as all the animals and plants certainly produced by

---
[1] "Lay Sermons," p. 226.
[2] *Ibid.*, p. 295.

selective breeding from a common stock are fertile with one another, that link will be wanting."[1]

"On a general survey of the theory," says Dr. Elam,[2] "nothing strikes us more forcibly than the total absence of direct evidence of any one of the steps. No one professes to have ever seen a variety (producing fertile offspring with other varieties) become a species (producing no offspring, or no fertile offspring, with the original stock). No one knows of any living or any extinct species having given origin to any other, at once or gradually. Not one instance is adduced of any variety having ever arisen which did actually give its possessor, individually, any advantage in the struggle for life. Not one instance is recorded of any given variety having been actually selected for preservation; whilst its allies became extinct. There is an abundance

---

[1] "Man's Place in Nature," p. 107.
[2] "Automatism and Evolution." *Contemporary Review*, vol. xxix. p. 131. [In gratefully acknowledging my indebtedness to the series of papers of which this is the third (for the first and second, see *Contemporary Review*, vol. xxviii. pp. 537 and 725), perhaps I may be permitted to say that, by their fairness and forcefulness, their clearness and conclusiveness, their breadth of range and their minuteness of detail, Dr. Elam has laid a large circle of readers under lasting obligations.]

of semi-acute reasoning upon what might possibly have occurred, under conditions which seem never to have been fulfilled;" but of direct and positive testimony, whether derived from the experience of mankind or from the geological record, there is no fragment whatever.

Mr. Darwin himself, as shown above,[1] is so far from pretending that his theory has received any "verification," as to acknowledge, with characteristic candour, that in the existence of structures which "cannot be accounted for by any form of selection,"[2] we have an objection which is "fatal" to that theory. And even in the case of other objections not thus pronounced absolutely "fatal" in form, his admissions are such as to show that they are fatal in fact. Thus, for instance, the absence of transitional forms between different species has always been recognised as a serious difficulty; and Mr. Darwin, in the attempt to obviate it, succeeds only in showing how very serious it is. These are his words:—

"Geology assuredly does not reveal any such finely graduated organic chain; and this, perhaps, is the most obvious and gravest objection which can be urged against

---

[1] *Ante*, p. 23.
[2] "Descent of Man," vol. ii. p. 387.

my theory. The explanation lies, as I believe, in the extreme imperfection of the geological record."[1]

But "the extreme imperfection of the geological record" here hypothecated by way of "explanation," is so far from being a scientific fact, that it was never imagined even by Mr. Darwin himself until he perceived that unless it were assumed, "the testimony of the rocks,"— not less than that of the "structures" presented by "man, as well as every other animal,"—would be "fatal" to his theory.

"I do not pretend that I should ever have suspected how poor a record of the mutations of life the best preserved geological section presented, had not the difficulty of our not discovering innumerable transitional links between the species which appeared at the commencement and close of each formation, pressed so hardly on my theory."[2] And again :—" He who rejects these views on the nature [3] of the geological record, will rightly reject my whole theory."[4]

On Mr. Darwin's own showing therefore, *cadit quæstio*. "These views" of his are to be rejected as unscientific, because they are unveri-

[1] "Origin of Species." Murray, 1859, p. 280.
[2] *Ibid.*, p. 302.
[3] *i.e.*, the alleged "extreme imperfection."
[4] "Origin of Species," p. 342.

fied. They are at best "a mere figment of the intellect." And their rejection involves the rejection of his "whole theory."

It is therefore no matter for surprise that a competent authority like Mr. St. George Mivart should conclude his exhaustive examination with these weighty words :—

"With regard to the conception as now put forward by Mr. Darwin, I cannot truly characterize it but by an epithet which I employ only with much reluctance. I weigh my words, and have present to my mind the many distinguished naturalists who have accepted the notion, and yet I cannot hesitate to call it a *'puerile hypothesis.'* "[1]*.

Mr. Mivart's judgments need no endorsement here; but those who are most conversant with the highly cultivated critical faculty, the profound knowledge of natural history and of biological science which in his "Genesis of Species," and afterwards, in his "Lessons from Nature," he has brought to the refutation of Mr. Darwin's doctrine of Natural Selection, will be the first to adopt and to reiterate this,

---

[1] "Lessons from Nature, as manifested in Mind and Matter." By St. George Mivart, Ph.D., F.R.S., etc. London: Murray, 1876. Chap. ix. p. 300. (* This emphasis of italics is Mr. Mivart's.)

"A Puerile Hypothesis." 61

his latest verdict. That doctrine lacks even the ordinary respectability of "a mere figment of the intellect." It is not merely fictitious, it is discreditable:—a "*puerile hypothesis.*"

CHAPTER IV.

"*SCIENTIFIC LEVITY.*"

AGNOSTIC Evolution, then, is merely an unverified hypothesis. And it is based on two subordinate hypotheses, equally unverified. And in relation to it, these last are so essentially necessary, so absolutely fundamental, that if either of them be invalidated the entire superstructure falls to the ground. The Evolution here controverted, has no existence whatever, has even no theoretical existence, apart from these two postulates: (1) "Spontaneous Generation"; and (2) "Transmutation of Species." Without the first, it would be destitute of its starting-point, the "primordial form." Without the second, it would still be destitute, on agnostic principles, of all other forms than one.

"Transmutation of Species," however, though reserved for further examination below, may for the present be dismissed, on the high authority of Professor Mivart, as a "puerile hypothesis." But when, on scientific grounds, we proceed to enquire as to the amount and character of evidence produced or producible, in favour of " Spontaneous Generation," we are compelled to regard it as a hypothesis still more puerile.

Speaking of evolution at large, and in comprehensive terms, Professor Whewell justly says,—" The system ought to be described as *a System of Order in which life grows out of dead matter, the higher out of the lower animals, and man out of brutes.*" [1]

To begin then at the beginning. Is "The System," in its first postulate, true or false ? Is it matter of fact, or merely matter of fiction? Does "life grow out of dead matter?"

Let us give the place of honour to "the Abraham of scientific men." Mr. Darwin, writing to the *Athenæum*, says—" I hope you will permit me to add a few remarks on Heterogeny, as the old doctrine of spontaneous generation is now called, to those given by Dr. Carpenter, who, however, is probably better fitted to discuss the question than any other man in England. Your reviewer believes that certain lowly organized animals have been generated spontaneously—that is, without pre-existing parents —during each geological period in slimy ooze. A mass of mud with matter decaying and under-

[1] Whewell's " Indications." Second Edition, p. 12.

going complex chemical changes is a fine hiding-place for obscurity of ideas. But let us face the problem boldly. He who believes that organic beings have been produced during each geological period from dead matter, must believe that the first being thus arose. There must have been a time when inorganic elements alone existed in our planet: let any assumptions be made, such as that the reeking atmosphere was charged with carbonic acid, nitrogenized compounds, phosphorus, etc. Now is there a fact, or a shadow of a fact, supporting the belief that these elements, without the presence of any organic compounds, and acted on only by known forces, could produce a living creature? At present, it is to us a result absolutely inconceivable."[1]

Dr. Carpenter had previously written thus:— "If your reviewer prefers to suppose that new types of Foraminifera originate from time to time out of the 'ooze,' under the influence of 'polar forces,' he has, of course, a right to his opinion; though by most naturalists such 'spontaneous generation' of rotalines and nummulites will be regarded as a far more 'astounding hypothesis' than the one for which it is offered as a substitute. But I hold that mine

[1] *The Athenæum* for 1863, p. 554.

is the more scientific, as being conformable to the fact . . . ; whilst his is not supported by any evidence that rotalines or nummulites ever originate spontaneously, either in 'ooze' or anywhere else."[1]

"Spontaneous generation" therefore, so far from being a scientific verity, is pronounced by the highest authority in England to be an "astounding hypothesis," " not supported by any evidence"; while the scientific Abraham declares it to be "absolutely inconceivable."

"What displeases me in Strauss," says Humboldt, "is the scientific levity which leads him to see no difficulty in the organic springing from the inorganic, nay, man himself from Chaldean mud."[2]

But how? The *modus operandi:* what was that? For answer we must turn first of all to a work which has at least the distinction of having obtained honourable mention by Prof. Tyndall. In the Belfast Address[3] we read of "the celebrated Lamarck, who produced so profound an impression on the public mind through the vigorous exposition of his views by the author of the 'Vestiges of Creation.'" Turning then

[1] *The Athenæum* for 1863, p. 461.
[2] "Letters to Varnhagen." First Edition, p. 117.
[3] P. 37.

to this "vigorous exposition" we find that the transition was effected by means of a "nucleated vesicle." This "nucleated vesicle," the fundamental form of all organisation, we must regard as "the meeting-point between the inorganic and the organic—the end of the mineral and the beginning of the vegetable and animal kingdoms, which thence start in different directions, but in a general parallelism and analogy."

Nor is this all. For "this nucleated vesicle is itself a type of mature and independent being in the infusory animalcules, as well as the starting-point in the fœtal progress of every higher individual in creation, both animal and vegetable."

What then? Granting all that is here assumed, we are as far as ever from a solution of the problem proposed. That problem is, to show the course of "Nature's great progression," as asserted, "from the formless to the formed, from the inorganic to the organic." But to begin with the nucleated vesicle as "the fundamental form of all organisation," is to begin, not at the beginning, but at the end. "The starting-point" here alleged, is on the wrong side the gulf. We want to know how it was reached. We want to see, not the first thing "formed," but the bridge that spans the chasm for the "great progression" from the formless; not the first thing that lived, but the "evolution" of "life" from "not living matter."

But to satisfy this demand is, as we have seen, impossible, since the "evolution" required is not only non-existent, but is pronounced by Mr. Darwin himself to be "absolutely inconceivable." What then is to be done? Nothing is more simple. The demand that cannot be met must be evaded; and we are accordingly asked to believe that the nucleated vesicle "is a form of being which there is some reason to believe electric agency will produce—though not perhaps usher into full life—in albumen, one of those component materials of animal bodies, in whose combination it is believed there is no chemical peculiarity forbidding their being any day realized in the laboratory. Remembering these things," proceeds the writer, "we are drawn on to the supposition that the first step in the creation of life upon this planet was a chemico-electric operation, by which simple germinal vesicles were produced."

Observe here, not the reasoning, but the unreason. The premiss, "There is some reason to believe." The conclusion, a "supposition." There is some reason to believe that "electric agency will produce" something not alive.

Ergo, "a chemico-electric operation" was "the first step in the creation of life!"

But had not Prevost and Dumas previously announced that "globules could be produced in albumen by electricity"? Quite true: but the support which the author's "supposition" was supposed to receive from that announcement fails at once before the remark that, "if his theory had been that the first step in the process of creation was the formation of vesicles by the wind passing over the ocean, then the fact of boys blowing bubbles in soap and water with a tobacco pipe, and the fable of Venus being born of the froth of the sea would have been as much to his purpose."

From the author of the "Vestiges" we turn to his eulogist, Professor Tyndall:—

"If you ask me whether there exists the least evidence to prove that any form of life can be developed out of matter, without demonstrable antecedent life, my reply is that evidence considered perfectly conclusive by many has been adduced; and that were some of us who have pondered this question to follow a very common example, and accept testimony because it falls in with our belief, we also should eagerly close with the evidence referred to. But there is in the true man of science a wish stronger than the wish to have his beliefs upheld; namely, the wish to have them true. And this stronger wish causes him to reject the most plausible support if he has reason to suspect that it is vitiated by error. Those

Scientific Sophisms. 69

to whom I refer as having studied this question, believing the evidence offered in favour of 'spontaneous generation' to be thus vitiated cannot accept it. They know full well that the chemist now prepares from inorganic matter a vast array of substances which were some time ago regarded as the sole products of vitality. They are intimately acquainted with the structural power of matter as evidenced in the phenomena of crystallization. They can justify scientifically their *belief* in its potency, under the proper conditions, to produce organisms. But in reply to your question they would frankly admit their inability to point to any satisfactory experimental proof that life can be developed save from demonstrable antecedent life. As already indicated, they draw the line from the highest organisms through lower ones down to the lowest, and it is the prolongation of this line by the intellect beyond the range of the senses that leads them to the conclusion which Bruno so boldly enunciated."[1]

Reserving, for the present, all consideration of the other important admissions in this remarkable paragraph, it is sufficient to note here the distinctly decisive answer which it furnishes to the question before us. "The evidence offered in favour of 'spontaneous generation'" is "vitiated by error." There is no "satisfactory experimental proof," nor even does there exist "the least evidence to prove that any form of life can be developed out of matter, without demonstrable antecedent life."

With this avowal of Professor Tyndall as

[1] "Belfast Address," pp. 55, 56.

well as with the preceding passage from the "Vestiges," it is instructive to compare the carefully constructed sentences—so reticent, so politic—of Mr. Herbert Spencer:—

"The chasm," he tells us, "between the inorganic and the organic is being filled up. On the one hand, some four or five thousand compounds, once regarded as exclusively organic, have now been produced artificially from inorganic matter; and chemists do not doubt their ability so to produce the highest forms of organic matter. On the other hand, the microscope has traced down organisms to simpler and simpler forms, until in the *Protogenes* of Professor Haeckel, there has been reached a type distinguishable from a fragment of albumen only by its finely granular character."[1]

On which Dr. Elam pertinently asks, "Does not every candid observer know that this said 'chasm' is not in any way 'being filled up;' and that the chemist could quite as easily construct a full-grown ostrich, as this despised bit of finely granulated albumen?" As for the "four or five thousand compounds," as well might the goldsmith say that he did not "doubt his ability" to make gold out of a

[1] "Principles of Psychology" (Stereotyped Edition), vol. i. p. 137.

Scientific Sophisms.

baser metal, because he had already moulded it and coloured it in four or five thousand different fashions. It is not in any sense true that any substance even distantly resembling organizable matter has been formed. The line of demarcation between the organic and the inorganic is as wide as ever. For what are these "organic" matters said to have been formed from their elements? They are chiefly binary and ternary compounds; certain acids of the compound radical class, some alcohols, ethers, and the like. Not one of them bears the most remote resemblance to anything that can live. Few of them contain nitrogen, and these few, chiefly *amides*, are only combinations of ammonia or ammonium with other binary or ternary compounds, and can only by courtesy or convention be allowed to be of "organic" nature. Neither chemically nor physically are they in any way allied to matter possessing the capacity of life. "One least particle of albumen, granulated or not granulated, would be an answer a thousandfold more crushing to the opponents of Evolution than myriads of such compounds."

It is now thirty-five years since the author of the "Vestiges," in his "vigorous exposition," enunciated the "belief" that "albumen" might be "any day realized in the laboratory;" and that there was "no chemical peculiarity forbidding" that realization. In those thirty-five years scientific chemistry has advanced, with colossal strides, at a rate of progress previously unknown and unimagined. Its triumphs are attested by the number and character of its investigations, its improved methods, its enlarged nomenclature, its ever-increasing wealth of results. Its history during the present century presents a continuous series of remarkable discoveries: the number of non-metallic elements has been increased by the addition of iodine, bromine, and selenium; that of the metals has been nearly doubled; the carefully examined compounds have increased a hundredfold; "a vast array of substances" has been compounded or decompounded; but, towards that border-land which separates the organic from the inorganic—if such a border-land there be—this triumphant chemistry has not advanced one single step. "Chemists," we are told, "do not doubt their ability" to do that which has hitherto mocked all their efforts. Thirty-five years ago they were equally untroubled by doubt, and equally destitute of achievement. They then believed the great desideratum might be "any day realized in the laboratory."

Scientific Sophisms. 73

And they "do not doubt" it now. But still they do not "realize" it. They have not "the least evidence" in support of their belief: they have still less of "satisfactory experimental proof."

But who is this "they"? It is not the chemist: it is the "philosopher." The chemist knows better. He knows that notwithstanding an altered classification of "organic" and "inorganic," yet between his compounds on the one hand, and the construction of *organizable* matter on the other, there still stands the impassable barrier which demonstrates that the affinities of life and living matter belong to a chemistry of which we know nothing, and which, to strive to imitate is but to strive in vain.

The name of Dr. Rudolf Virchow has been familiar to scientific Europe for nearly forty years, as one honoured amongst the most honourable. It was he who, at the Conference of the Association of German Naturalists and Physicians at Munich, in the autumn of 1877, led the reaction in the high places of learning against the dogmatism of science. And this is what he says on the "scientific levity" of "spontaneous generation":—

"I grant that if any one is determined

to form for himself an idea of how the first organic being could have come into existence, *of itself*, nothing further is left than to go back to spontaneous generation. . . . But of this we do not possess any actual proof. No one has ever seen a *generatio æquivoca* really effected; and whoever supposes that it has occurred is contradicted by the naturalist, and not merely by the theologian. . . . If it were capable of proof, it would indeed be beautiful! . . . But whoever recalls to mind the lamentable failure of all the attempts made very recently to discover a decided support for the *generatio æquivoca* in the lower forms of transition from the inorganic to the organic world, will feel it doubly serious to demand that this theory, so utterly discredited, should be in any way accepted as the basis of all our views of life."[1]

An "astounding hypothesis," "not supported by any evidence,"[2] "absolutely inconceivable,"[3] and "utterly discredited."[4] Such is the "scientific levity" of Spontaneous Generation.

[1] "The Freedom of Science in the Modern State," p. 39.
[2] Dr. Carpenter, *ut sup.*
[3] Mr. Darwin.
[4] Dr. Virchow.

CHAPTER V.

A HOUSE OF CARDS.

"SPONTANEOUS Generation" therefore, not less than "Transmutation of Species," is merely "a puerile hypothesis." But on these two dogmas the theory of agnostic Evolution is absolutely dependent. By means of the support derived from them—if only they themselves could have been made to stand—it might have stood; but with their fall, it also comes to the ground. Its relation to them renders its fate inevitable. The instability of the superstructure is inseparably concomitant with the insecurity of the foundation.

Nor is this all. Fate is involved in character: and when we proceed to examine the character of this theory, we are at no loss to discover the cause of its fate.

If the doctrine of agnostic Evolution were scientifically true, it could not fail to command the universal assent of scientific men; whereas now, on the contrary, it is notorious that among the ranks of those most eminent for scientific attainment there are not wanting earnest and enlightened seekers after truth, who have not only refused to accept this new doctrine with its "logical consequences," but who have based their refusal on this explicit ground, that agnostic Evolution is "nothing more than a flimsy framework of hypothesis constructed upon imaginary or irrelevant facts, with a complete departure from every established canon of scientific investigation."

In his Review of Professor Haeckel's "Natural History of Creation," or, as he would prefer to call it, "The History of the Development or Evolution of Nature," Professor Huxley has expressly formulated "the fundamental proposition of Evolution." "That proposition is," he tells us, "that the whole world, living and not living, is the result of the mutual interaction, according to definite laws, of the forces possessed by the molecules of which the primitive nebulosity of the universe was composed."[1] And he adds, "If this be true, it is no less certain that the existing world lay potentially in the cosmic vapour."

In this, of course, he agrees with Haeckel, by

[1] "Critiques and Addresses," Macmillan, 1873 (xii. "The Genealogy of Animals"), p. 305.

A House of Cards. 77

whom "full justice is done to Kant, as the originator of that 'cosmic gas theory,' as the Germans somewhat quaintly call it, which is commonly ascribed to Laplace."[1]

Professor Tyndall agrees with both. Having discerned in "Matter" "the promise and potency of all terrestial life,"[2] he lays it down as fundamental that "the doctrine of evolution derives man in his totality from the interaction of organism and environment through countless ages past."[3] By that "vision of the mind," which, as he tells us, "authoritatively supplements the vision of the eye,"[4] he sees "the cosmic vapour" as a primitive "nebular haze" (the "universal fire-mist" of the "Vestiges"), gradually cooling, and contracting as it cooled, into a "molten mass," in which "latent and potential" were not only "life" before it was alive, and "form" before it was formed, "not alone the exquisite and wonderful mechanism of the human body, but the human mind itself; emotion, intellect, will, and all their phenomena . . . all our philosophy, all our poetry, all

---

[1] "Critiques and Addresses," Macmillan, 1873 (xii. "The Genealogy of Animals"), p. 304.
[2] "Belfast Address," p. 55.
[3] *Ibid.*, p. 59.
[4] *Ibid.*, p. 55.

our science, and all our art—Plato, Shakespeare, Newton, Raphael." All that has been; all that is; nay even all that is imagined only; was once,—to the scientific eye, "in a fine frenzy rolling,"—"potential in the fires of the sun;"[1] just as those fires themselves had no existence (other than "latent and potential") until they were kindled by the condensation of "the cosmic vapour."

These quotations, and such as these—for they might be indefinitely extended—enable us to sum up the doctrine of Agnostic Evolution in two short propositions :—

First, " That the earliest organisms were the natural product of the interaction of ordinary inorganic matter and force."

Second, "That all the forms of animal and vegetable life, including man himself, with all his special and distinctive faculties, have been slowly, but successively and gradually developed from the earliest and simplest organisms."

But when we proceed to examine the scientific pretensions of the theory thus succinctly stated, we find, on Professor Tyndall's own showing, that they are worthless. Worthless, because unverified, and incapable of verification. "The strength of the doctrine of

[1] "Scientific Use of the Imagination," p. 453.

evolution consists," he tells us, "not in an experimental demonstration (for the subject is hardly accessible to this mode of proof), but in its general harmony with scientific thought."[1] "Scientific thought," however, can only mean "the aggregate thoughts of scientific men;" and with these thoughts it is most certain that this doctrine of Evolution is *not* in harmony. Mr. Darwin, with his usual candour, writes as recently as 1871, "Of the older and honoured chiefs in natural science, many unfortunately are still opposed to Evolution in every form."[2] Since that date it is certain that, on the continent at least, the doctrine has been met by many distinguished botanists and zoologists with growing disfavour. To the same purpose is the still more recent admission of Professor Tyndall: "Our foes are to some extent they of our own household, including not only the ignorant and the passionate, but a minority of minds of high calibre and culture, lovers of freedom, moreover, who, though its objective hull be riddled by logic, still find the ethic life of their religion unimpaired."[3]

But even if this were not the case, it would

[1] "Belfast Address," p. 58.
[2] "Descent of Man," p. 2.
[3] "Materialism and its Opponents," *ut sup.*, p. 597.

still be true, on Professor Tyndall's showing, that Evolution as above defined has not been "verified" "by observation and experiment;" and that "without verification a theoretic conception is a mere figment of the intellect."[1] "Those who hold the doctrine of evolution," he tells us, "are by no means ignorant of the uncertainty of their data, and they only yield to it a provisional assent. They regard the nebular hypothesis as probable, . . . and accept as probable the unbroken sequence of development from the nebula to the present time."[2]

"Probable," "provisional," "uncertain," and therefore "unscientific;" this, on the highest authority, is thus admitted to be the actual character of "the doctrine of Evolution." But of what kind is this probability? When examined, it appears that even the alleged probability itself is at best a mere "supposition," "a theoretic conception," a probability hypothetical only, nothing more.

For example: Mr. Herbert Spencer tells us that "there is reason to suspect that there is but one ultimate form of Matter, out of which the successively more complex forms of Matter

---

[1] "Fragments of Science," p. 469.
[2] "Scientific Use of the Imagination," p. 456.

A House of Cards. 81

are built up."[1] When we ask for the reason for this assertion, we are merely told that there is "reason to suspect" so, and that "by the different grouping of units, and by the combination of the unlike groups each with its own kind, and each with other kinds, it is supposed that there have been produced the kinds of matter we call elementary."[2] But, for anything that appears to the contrary, the "reason to suppose" all this, and the subsequent supposing of it, exist only in Mr. Spencer's own mind, and have their *raison d'être* in the exigencies of the "constructive philosophy." Having however in this way "supposed" whatever he pleased, and having also justified his method of procedure by saying that there was "reason to suppose" so, he then in the very next paragraph, and without adducing any proof whatever, proceeds to treat these suppositions as if they were ascertained facts, and builds on them as if he took them for solid foundations. Thus :—" If then, WE SEE (!) that by unlike arrangements of like units, all the forms of matter, apparently so diverse in nature may be produced," etc. etc.[3]

---

[1] "Principles of Psychology." Stereotyped Edition. Williams & Norgate, 1870, vol. i. p. 155.
[2] *Ibid.*   [3] *Ibid.*

But this method of evolving science from supposition, and conjuring with conjecture for certainty, is by no means a monopoly of Mr. Herbert Spencer. In one sentence of his Essay on "Scientific Materialism," Professor Tyndall states that "we should on philosophic grounds *expect to find*" certain physical conditions; and in the next, he commences an induction, from this mere expectation, with the phrase, "The relation of physics to consciousness *being thus* invariable"![1] a relation which, if it exists at all, does certainly not exist in any demonstrable form— so far is it from "being thus," or being in any way other than that of "expectation" merely, "invariable."

Similarly, when, in his controversy with Mr. Martineau, he claims "consciousness" for the fern and the oak, he says, "No man can say that the feelings of the animal are not represented by a drowsier consciousness in the vegetable world. At all events no line has ever been drawn between the conscious and the unconscious; for the vegetable shades into the animal by such fine gradations, that it is impossible to say where the one ends and the other begins. . . . The evidences as to

[1] "Fragments of Science." Sixth Edition. Longmans, 1879, vol. ii. p. 86.

A House of Cards. 83

consciousness in the vegetable world depend wholly upon our capacity to observe and weigh them."[1] What then? This, evidently: that since we are not possessed of any such capacity; and since, without that capacity the evidence is non-existent; it follows that there is no evidence whatever "as to consciousness in the vegetable world." But if there is a fatal lack of evidence there is no lack of imagination; and Dr. Tyndall's imagination, always brilliant, is fully equal to the occasion. He supposes altered conditions for the observer, and then says: "I can imagine not only the vegetable, but the mineral world, responsive to the proper irritants."[2] "I can imagine!" What? "Consciousness" in a cabbage, and in a granite cube. But on what evidence? None that I can find: but plenty that "I can imagine!"

In the same category with the suppositions of Mr. Spencer and the imagination of Professor Tyndall must be placed the conceptions of Mr. Darwin. Like them, he has to assume as fundamental, certain propositions which he cannot prove. But then if he cannot prove, he "cannot doubt," or he "can hardly doubt;" and this incapacity for doubt serves as a highly

[1] "Materialism and its Opponents," *ut sup.*, p. 595.
[2] *Ibid.*

convenient substitute for certainty. Thus, *e.g.*—

"I cannot doubt that the theory of descent with modification embraces all the members of the same class."[1] And again: "I can indeed hardly doubt that all vertebrate animals having true lungs, have descended, by ordinary generation from an ancient prototype, of which we know nothing, furnished with a floating apparatus or swim-bladder." " It is conceivable that the now utterly lost branchiæ might have been gradually worked in by natural selection for some quite distinct purpose, in the same manner as . . it is probable that organs which at a very ancient period served for respiration, have been actually converted into organs of flight."[2]

It would be sufficiently surprising, if we had not been so long accustomed to it, to learn that the possession of lungs which constitute the fitness of the possessors for living, not in water, but in air, betrays their aquatic origin.[3] But it is much more surprising that men illustrious in virtue of their scientific eminence should expect

[1] "Origin of Species," p. 484.
[2] *Ibid.*, p. 191.
[3] "Land animals, which in their lungs or modified swim-bladders betray their aquatic origin." (*Ibid.*, p. 196.)

A House of Cards. 85

a tissue of conjectures such as this to be accepted as if it possessed any scientific authority.

The branchiæ are "now utterly lost;" that is, they are non-existent, except to the "imagination," to which "it is conceivable" that they might once have been otherwise. That "ancient mariner," the primeval ancestor of the human race, was "an ancient prototype of which we know nothing." And yet, strange to say, we do know this: that he was "furnished with a floating apparatus or swim-bladder." Some thing "might have been" made of the missing branchiæ "for some quite distinct purpose;" for this, although not actual is at least "conceivable." Nay, it almost emerges from the realm of the ideal when we are to be shown the *modus operandi*,—"in the same manner as"—as what? As in some other instance of which we have tangible proof? No, not that: but only as in some other instance where "it is probable," or at least supposable, that "organs which at a very ancient period" may or may not have existed to serve a given end, would be of great service to this theory if only it could be shown first, that they did exist, and then that they ceased to exist, by having been "actually converted" into other organs to serve another and a very different end.

Mr. Spencer "supposes;" Dr. Tyndall "imagines;" Mr. Darwin "conceives." Tier on tier the towering fabric totters to its fall: no stability in the foundation, no continuity in the superstructure; "a flimsy framework of hypothesis, constructed upon imaginary or irrelevant facts, with a complete departure from every established canon of scientific investigation."

CHAPTER VI.

SOPHISMS.

"No stability in the foundation, no continuity in the superstructure"; "a flimsy framework of hypothesis, constructed on imaginary facts." If any one imagines that this is the language of exaggeration or romance, let him turn to the twenty-second chapter of Haeckel's "Natural History of Creation," where he will find a complete and circumstantial history of human ancestry in twenty-two stages of existence, from the unicellular *Monera* up to the perfect man. The theory of man's ape-descent thus constructed is perfect—but it is in the air. It lacks but one thing to give it relevance : and that one thing is reality. Like the "châteaux en Espagne" of the penniless Count, it exists only in the interested imagination of the pretender.

Du Bois Reymond has incurred the bitter and unappeasable wrath of Haeckel by declaring this genealogical tree (*Stammbaum*) to be as authentic in the eyes of a naturalist, as are the pedigrees of the Homeric heroes in those of an historian. And no wonder; for, unauthentic and unreal as they are, they are indispensable. Without them the theory of evolution has no pretence to "continuity." But with their aid, although the continuity which they confer is still *in nubibus*, the argument is rounded with the completeness of a circle. What are the proofs of man's descent from the ape? The facts of ontogenesis[1] and phylogenesis and their correspondence. Where are these facts enumerated? In the twenty-second chapter of Haeckel's "Natural History of Creation." What is the authority for these facts? Chiefly this: that they are necessitated by the exigencies of the theory. But where is the demonstrative evidence, direct or indirect, that any creatures representing these twenty-two imaginary stages ever existed? In the majority of instances there is no such evidence; but they "must have existed," otherwise the theory would be imperfect.

For example, the *Monera*, according to Haeckel, are our earliest "ancestors;" and of these it is stated,—as if it were a plain historical

---

[1] *Ontogenesis*, the history of individual development.
Phylogenesis, the history of genealogical development.
Biogenesis, the history of life development generally. [aeckel.)

Sophisms. 89

fact,—that "they originated about the beginning of the Laurentian period, by *archebiosis* or spontaneous generation," from "so-called inorganic compounds of carbon, hydrogen, oxygen, and nitrogen."[1] After what has been already said of spontaneous generation,[2] it is almost superfluous to add that this assertion about our earliest "ancestors" is not only destitute, it is also incapable, of proof. And yet the fundamental law (*Grundgesetz*) of ontogenesis absolutely requires it.

Again. In his Munich Address, Haeckel repeats the trite old tale ("as if it had not been a hundred times blown into the 'infinite azure'") that "the Monera, consisting of protoplasm only, bridge over the deep chasm between organic and inorganic nature, and show us *how* the simplest and oldest organisms *must have* originated from inorganic carbon compounds."[3] Whereas, on the contrary, the simple fact is that the Monera *bridge over nothing whatever;* nor do they show, in any conceivable way, *how* life has originated from inorganic compounds. Chemically and dynamically the protoplasm of

[1] "Naturliche Schöpfungsgeschichte," p. 578.
[2] *Vide sup.*, p. 50 et seqq., especially p. 59.
[3] "Die Heutige Entwickelungslehre im Verhältnisse zur Gesammtwissenschaft," p. 13.

these apparently simple organisms is just as far removed from inorganic matter as is the protoplasm of the lion or the eagle.

Of another important group of "ancestors," the *Gastreada*, we are told that it "*must have* existed in the primordial time, and must have included amongst its members the direct ancestors of man." No one ever saw a single individual of this group; that is a matter of course. It is equally a matter of course that no traces are to be found of its existence. But the "certain proof"[1] of that existence is supposed to be found in the fact that the Amphioxus, at one period of its development, presents a type similar to that of—of what? Of the imaginary Gastræa whose existence had to be proved! Our ancestors, the worms, come next; and, like their predecessors, they "*must have* existed," because without them we should be at a loss how to derive higher worms, and the articulata generally.

Professor Huxley, summarizing and reviewing this volume of Haeckel's, is careful to express his "entire concurrence with the general tenor and spirit of the work," and his "high estimate of its value."[2] Of the particular por-

[1] "Naturliche Schöpfungsgeschichte," p. 581.
[2] "Critiques and Addresses." Macmillan, 1873, p. 319.

Sophisms. 91

tion now under review, he says,—"In Professor Haeckel's speculation on Phylogeny, or the genealogy of animal forms, there is much that is profoundly interesting, and his suggestions are always supported by sound knowledge and great ingenuity. Whether one agrees or disagrees with him, one feels that he has forced the mind into lines of thought in which it is more profitable to go wrong than to stand still.

"To put his views into a few words, he conceives that all forms of life originally commenced as *Monera*, or simple particles of protoplasm; and that these Monera originated from not living matter. Some of the Monera acquired tendencies towards the Protistic, others towards the Vegetal, and others towards the Animal modes of life. The last became animal *Monera*. Some of the animal *Monera* acquired a nucleus, and became amœba-like creatures; and out of certain of these, ciliated infusorium-like animals were developed. . These became modified into two stirpes: A, that of the worms; and B, that of the sponges. The latter by progressive modification gave rise to all the *Cœlenterata;* the former to all other animals. But A soon broke up into two principal stirpes, of which one, *a*, became the root of the *Anne-*

G

lida, *Echinodermata*, and *Arthropoda*, while the other, *b*, gave rise to the *Polyzoa* and *Ascidioida*, and produced the two remaining stirpes of the Vertebrata and the Mollusca."[1]

Many persons will agree with Mr. Huxley so far as to admit that Professor Haeckel is not destitute either of "sound knowledge," or of "great ingenuity," who yet think Mr. Huxley in error when he represents his favourite Professor as possessing these characteristics in combination. As displayed in his "speculations on Phylogeny," they appear to be not so much in combination as in opposition. Each invades the province of the other. Take away the "knowledge," and you clear the field for the "ingenuity": but where "sound knowledge" is supreme, "great ingenuity" is superfluous. He who finds it "more profitable to go wrong than to stand still," may indeed display "great ingenuity," but the soundness of his "knowledge" is by no means unquestionable.

Take, for example, this very summary of "his views," as here given by Professor Huxley. What he does "view" is something not actual and real, but ideal only. He does not "prove"; he does not even assign reasons for belief; but,

[1] "Critiques and Addresses." Macmillan, 1873, pp. 314, 315.

like Mr. Darwin, he merely "conceives" a certain ideal origin of life. His *Monera*, at first "conceivable" only, and then "conceived," "acquired tendencies." But how did they acquire them? And how does he know that they were acquired? The only answer is, that they *must have* acquired them or they could never have possessed them and they *must have* possessed them, or they could not have become animal *Monera;* and they *must have* become animal *Monera,* for without them the theory breaks down, and the existence of the animal world could be accounted for only by admitting the doctrine of a special creation. To meet the exigencies of the theory therefore, these "simple particles," so inexplicably "originated," and with "tendencies" so inexplicably "acquired," at last, and in some equally inexplicable manner, "*became* animal *Monera.*"

"At last!" By no means: this is but another beginning. Each tier of the hypothesis is constructed only by a recurrence of the same dogmatic assumptions. "Some of the animal *Monera* acquired a nucleus, and became amœba-like creatures." "Great ingenuity?" Undoubtedly: whatever the theory requires is forthcoming——on paper. The transformations are as surprising, as unaccountable,—and as unreal,—as those which ingenuity, by means of sleight of hand, brings out of a conjuror's hat. But it is only conjuring after all; and "sound knowledge" is not imposed upon by sleight of hand. These "simple particles" "originated," "acquired," "became," "were developed," "became modified," "gave rise to," and "produced," "all forms of life." How? When? Where? No such origination has ever been witnessed. No such evolution has ever been observed. No such results have ever been produced. But the theory requires them; and consequently, to meet the exigencies of the theory, here they are —on paper.

Before dismissing "Professor Haeckel's speculations on Phylogeny," there is one other point that calls for special notice. His fundamental postulates are these: "That all forms of life originally commenced as *Monera*, or simple particles of protoplasm; and that these *Monera* originated from not living matter." Yet he himself is perfectly aware that these, his fundamental postulates, are not only "not proven," but are incapable of proof. "With respect to spontaneous generation," says Mr. Huxley,[1] "while admitting that there is no experimental

[1] "Critiques and Addresses." Macmillan, 1873, p. 304.

Sophisms. 95

evidence in its favour, Professor Haeckel denies the possibility of disproving it, and points out that the assumption that it has occurred is a necessary part of the doctrine of evolution." So be it. A more complete confirmation of what has been already said on this subject it would be impossible to desire. Evolution now, of necessity, rests on "spontaneous generation:" while spontaneous generation is at best an "assumption" of which its most uncompromising advocate admits that "there is no experimental evidence in its favour." So much the worse for "the doctrine of Evolution."

The position assumed by Mr. Huxley himself in reference to this subject is peculiar; so peculiar, indeed, that it had better be stated in his own words. In his Presidential Address to the British Association for the Advancement of Science (1870), he discusses the conflicting claims of *Biogenesis and Abiogenesis*, in one of the ablest and most lucid expositions ever given of that problem. By the former term he denotes "the hypothesis that living matter always arises by the agency of pre-existing living matter;" the latter term denotes the contrary doctrine—that living matter may be produced by matter not living.

The first distinct enunciation of the hypothesis that all living matter has sprung from pre-existing living matter, he traces not to our great countryman, Harvey, but to a contemporary though a junior of Harvey, and trained in the same schools, Francesco Redi. And he concludes his sketch of the progress of the doctrine, and of the successive experiments by which its truth has been established, in these words: "So much for the history of the progress of Redi's great doctrine of Biogenesis, which appears to me, with the limitations I have expressed, to be victorious along the whole line at the present day."[1]

His own adhesion to this "great doctrine of Biogenesis" is thus stated: "If in the present state of science the alternative is offered us,—either germs can stand a greater heat than has been supposed, or the molecules of dead matter, for no valid or intelligible reason that is assigned, are able to rearrange themselves into living bodies, exactly such as can be demonstrated to be frequently produced another way,—I cannot understand how choice can be, even for a moment, doubtful.

"But though I cannot express this conviction of mine too strongly, I must carefully guard myself against the supposition that I intend to

[1] "Critiques and Addresses," p. 239.

suggest that no such thing as Abiogenesis ever has taken place in the past, or ever will take place in the future. With organic chemistry, molecular physics, and physiology yet in their infancy, and every day making prodigious strides, I think it would be the height of presumption for any man to say that the conditions under which matter assumes the properties we call 'vital' may not some day be artificially brought together. All I feel justified in affirming is, that I see no reason for believing that the feat has been performed yet."[1]

Analysing this declaration we have three several propositions. Spontaneous generation is a dogma for which "no valid or intelligible reason is assigned." As between life derived from antecedent life, and life derived from something that was not alive, Professor Huxley "cannot understand how choice can be, even for a moment, doubtful." And "this conviction" of his he "cannot express too strongly." At the same time, however, he is not quite sure that the opposite of all this may not be also true—of some possible future, or perhaps even of some actual past.

But the climax is yet to come. The declaration above quoted,—"All I feel justified in af-

[1] "Critiques and Addresses," p. 238.

firming is, that I see no reason for believing that the feat has been performed yet,"—rests on reasons at once valid and intelligible, assignable and assigned. Any declaration, therefore, antagonistic to this, must of necessity be devoid of reason. Yet such is precisely the declaration which, in the very next paragraph, Professor Huxley proceeds to make. "If it were given me," he says, "to look beyond the abyss of geologically-recorded time . . . I should expect to be a witness of the evolution of living protoplasm from not living matter."[1] He would "expect to witness," in that "remote period," the performance of a feat which he sees "no reason for believing" has ever "been performed yet."

Professor Tyndall believes that if a planet were "carved from the sun, set spinning round an axis, and revolving round the sun at a distance from him equal to that of our earth,"[2] one of the "consequences of its refrigeration" would be "the development of organic forms." If you ask what reason can be assigned for this belief, you are asked in turn, "Who will set limits to the possible play of molecules in a cooling planet?"[3]

This conclusive question is suggestive of

[1] "Critiques and Addresses," p. 239.
[2] "Fragments of Science." Sixth Edition (1879), vol. ii. p. 51.
[3] *Ibid.*

Sophisms.

another :—" Who will set limits to the possible play of Professor Tyndall's scientific imagination ?" Why should a cooling planet be so much more likely to produce minute organisms, and to develope "organic forms," than a cooling flask? Or, as Dr. Bastian pertinently puts it, "If such synthetic processes took place then, why should they not take place now? Why should the inherent molecular properties of various kinds of matter have undergone so much alteration?"[1]

The opening sentences of the Belfast Address are vitiated by a fallacy which reappears in other places with the regularity of a recurring decimal. "An impulse inherent in primeval man," says Dr. Tyndall, "turned his thoughts and questionings betimes towards the sources of natural phenomena. The same impulse, inherited and intensified, is the spur of scientific action to-day. Determined by it, by a process of abstraction from experience we form physical theories which lie beyond the pale of experience, but which satisfy the desire of the mind to see every natural occurrence resting upon a cause."

Now, since of this "primeval man" nothing whatever is known, on what ground can it be affirmed that he possessed the "inherent impulse" here attributed to him? All that *is*

[1] "Beginnings of Life," Pref. p. x.

known of him is that his "progenitors" "could be not called human."[1] How came he then by this "inherent" impulse—an impulse now "inherited" as the distinctive characteristic of all mankind—yet not possessed by his non-human ancestors, and therefore not derived from them?

Inexplicable however as is this impulse, it is as nothing when compared with the theories to which it has given rise. The theories have been invented to satisfy a desire of the mind: the desire "to see every natural occurrence resting upon a cause." And to satisfy this desire the scientific imagination of to-day forms "physical theories which lie beyond the pale of experience," and rest—upon nothing. If, as the same eminent authority has told us, a "theoretic conception" is a mere "intellectual figment," until it has been "verified" by "observation and experiment," how is it possible that "theories which lie beyond the pale of experience," should satisfy a mind that desires "to *see* every natural occurrence resting upon a cause"? "Physical theories," to be satisfactory to such a mind, must lie within—and not beyond—the pale of experience.

"The porter sits down on the weight which he bore,"

---

[1] Professor Tyndall's (Birmingham Address) "Science and Man," p. 611.

says Wordsworth. And this he may do with perfect safety, even on the parapet of London Bridge; for that is within the pale of experience. But woe to the unlucky wight who, in the attempt to satisfy his desire for rest, ventures to sit down on some "abstraction" outside the parapet; for that is "beyond the pale of experience."

"Trace the line of life backwards," says our Lucretian, "and see it approaching more and more to what we call the purely physical condition. . . . We break a magnet and find two poles in each of its fragments. We continue the process of breaking; but, however small the parts, each carries with it, though enfeebled, the polarity of the whole. And when we can break no longer, we prolong the intellectual vision to the polar molecules. Are we not urged to do *something* similar in the case of life? . . . Believing as I do in the continuity of Nature, I cannot stop abruptly where our microscopes cease to be of use. Here the vision of the mind authoritatively supplements the vision of the eye. By an intellectual necessity I cross the boundary of the experimental evidence, and discern in Matter . . . the promise and potency of all terrestrial Life."[1]

[1] "Belfast Address," p. 55.

This "potency" of matter, then, when discerned at all, is discerned only "beyond the pale of experience," "across the boundary of experimental evidence." Scientifically, therefore, it is non-existent; a mere "intellectual figment," the product of an imaginary "intellectual necessity": an "unverified theoretic conception," nothing more; and this only when it has been actually "discerned." But, as simple matter of fact, it has never yet been actually discerned. Professor Tyndall himself has not thus discerned it. What he here calls discernment he elsewhere calls the scientific use of the Imagination. It is he himself who warrants the affirmation that this alleged "potency of all terrestrial Life" has not been discerned in Matter at all; it has only been imagined. "Conscious life" is a part, and the principal part, of "all terrestrial life." Has the life of a fern or an oak this potential "consciousness"? It is Dr. Tyndall who answers, "No man can tell."[1] Does pig iron possess this potency of conscious cogitation? or does the loftiest granite needle of the Alps cheer its eternal solitude with the reflection, "*Cogito, ergo sum*"? There

---

[1] "Materialism and its Opponents," *Fortnightly Review*, vol. xviii. p. 595. "Fragments of Science," Introduction.

is no answer. They make no sign. No such promise or potency is exhibited, and it is therefore no wonder that it is not discerned. But alter the conditions of discernment, says Dr. Tyndall, and then "I can imagine not only the vegetable, but the mineral world, responsive to the proper irritants."[1] Not, "I have discerned"; nor even I *can* discern; but only "I can *imagine!*"

And here the matter might be left, were it not that Dr. Tyndall has himself compelled us to ask whether he has not estimated too highly his own power of imagination. For how can even he imagine that which he himself tells us is unimaginable? The passage from physics to consciousness, he tells us,[2] "is unthinkable." "You cannot satisfy the human understanding in its demand for logical continuity between molecular processes and the phenomena of consciousness. This is a rock on which materialism must inevitably split whenever it pretends to be a complete philosophy of life."[3] Nor would the result be altered if even the experiment could be made under the altered conditions

---

[1] "Materialism and its Opponents," *Fortnightly Review*, vol. xviii. p. 595. "Fragments of Science," Introduction.

[2] *Ibid*, p. 589. [3] "Belfast Address," p. 33.

which in the passage above cited, it was found necessary to hypothecate. "Alter the capacity" of the observer, it was then said, "and the evidence would alter too."[1] Yet here, only six pages earlier, in the very same paper, we are told: "Were our minds and senses so expanded, strengthened, and illuminated, as to enable us to see and feel the very molecules of the brain; were we capable of following all their motions, all their groupings, all their electric discharges, if such there be; and were we intimately acquainted with the corresponding states of thought and feeling, we should be as far as ever from the solution of the problem, 'How are these physical processes connected with the facts of consciousness?' The chasm between the two classes of phenomena would still remain intellectually impassable."[2]

Yet notwithstanding all this Dr Tyndall formally proclaims his "belief" "in the *continuity* of Nature." The "continuity" of an "impassable chasm"! A chasm "intellectually impassable"; and yet "by an intellectual necessity" he crosses it. "Two classes of phenomena," and no possible means of transition

[1] "Materialism and its Opponents," p. 595.
[2] *Ibid.*, p. 589.

from one to the other. For, in order to "discern in matter the *promise*" of conscious life, we must be able,' by observation of its merely physical movements, to forecast, in a world as yet insentient, the future phenomena of thought and feeling. Yet this is precisely the transition which is pronounced "unthinkable." "We do not possess the intellectual organ, nor apparently any rudiment of the organ, which would enable us to pass, by a process of reasoning, from the one to the other. They appear together, but *we do not know why*."[1]

It is an instructive spectacle. Professor Huxley "expecting" to witness, in the remote past, the performance of a feat which he sees "no reason for believing" has ever yet been performed; and Professor Tyndall, "by an intellectual necessity" and a "vision of the mind," crossing "the chasm" "intellectually impassable" which separates two classes of phenomena, although he does "not possess the intellectual organ, nor apparently any rudiment of the organ, which would enable him to pass, by a process of reasoning, from the one to the other."

[1] "Materialism and its Opponents," p. 589.

Horace was undoubtedly right :—

". . . quandoque bonus dormitat Homerus."[1]

But had he lived in our time, and written of the Homers of modern materialism; had he heard their conjectural hypotheses, their conflicting asseverations, their autocratic dogmatism ;—

"Matter is the origin of all that exists; all natural and mental forces are inherent in it":[2] "All the natural bodies with which we are acquainted are *equally living: the distinction which has been held as existing between the living and the dead does not really exist:*"[3] "The eternal is the nothing of nature :" "There is no other science than that which treats of nothing :"[4] "Holothuriæ engender snails ;"[5] and "gazing upon a snail, one believes that he finds the prophesying goddess sitting upon the tripod ;" for "a snail is an exalted symbol of mind, slumbering deeply within itself :"[6] while

---

[1] "Ars Poet.," 359.
[2] Buchner's "Kraft und Stoff." (Collingwood's Translation) p. 32.
[3] "Naturliche Schöpfungsgeschichte." By Dr. Ernst Haeckel. Sixth Edition, p. 21.
[4] "Physiophilosophy" of Prof. Oken.
[5] Buchner's "Kraft und Stoff," p. 80.
[6] Oken.

Sophisms.

"Self-consciousness is a living ellipse:"[1] and Man is merely an automaton, though "a conscious automaton;" an "automaton endowed with free will:"[2]—

Had Horace heard all this, he would have had something more to say about this snail-like "mind, slumbering deeply;" and would have used a much stronger word than "quandoque."

[1] Oken.
[2] Prof. Huxley, in *The Fortnightly Review*, November, 1874, p. 577.

CHAPTER VII.

PROTOPLASM.

THE word "protoplasm" was invented in the year 1846, by the eminent German botanist Von Mohl, as a name for one portion of those nitrogenous contents of the cells of living plants, the close chemical resemblance of which to the essential constituents of living animals had been in that same year, emphatically pointed out by the botanist Payen. But if, pushing our investigation beyond the origin of the name, we inquire as to the nature of the thing, and ask What is Protoplasm? the answer to that question involves a reference to the historical progress of the physiological cell theory.

That theory may be said to have wholly grown up since Dr. John Hunter wrote his celebrated work "On the Nature of the Blood." According to Dr. Hunter, new growths depended on an exudation of the *plasma* of the blood, in which, by virtue of its own *plasticity*, vessels formed, and conditioned the further progress. When, at a later date, the conception of a cell had been arrived at, Schleiden, for starting point, required an intracellular plasma, and Schwann, a structureless exudation, in which minute granules, if not indeed already pre-existent, formed, and by aggregation grew into nuclei, round which singly the production of a membrane at length enclosed a cell. Brown demonstrated a nucleus in the vegetable cell; as Valentin subsequently did in the animal one; Müller insisted on the analogy between animal and vegetable tissue; Schwann's labour in completing the theory of the animal cell may be regarded as completing the first stage of the cell theory: but the raising it to the second stage must be attributed to the wonderful ability of Virchow. And it is to the resolution of this second stage that we owe the word Protoplasm.

In Virchow's view, the body constituted a free state of individual subjects, with equal rights but unequal capacities. These were the cells, which consisted each of an enclosing membrane, and an enclosed nucleus with surrounding intracellular matrix or matter. These cells propagated themselves, chiefly by partition or division; and the fundamental principle of the entire theory was expressed in the dictum, "*Omnis cellula e cellula.*"

The first step in resolution of this theory was the elimination of the investing membrane. Such membrane may and does ultimately form; but in the first instance, for the most part, the cell is naked. The second step was the elimination, or at least the subordination, of the nucleus. The nucleus is now discovered to be necessary neither to the division nor to the existence of the cell.

Thus, then, stripped of its membrane, relieved of its nucleus, what now remains for the cell? Nothing, but that which *was* the contained matter, the intracellular matrix, and *is*—Protoplasm.

The application of the word, however, to the element in question, like the history of the thing, was marked by several stages. First came Dujardin's discovery of sarcode. Then, as above mentioned, Von Mohl's introduction of the term protoplasm as the name for the layer of the *vegetable* cell that lined the cellulose, and enclosed the nucleus. Cohn, four years later, proclaimed "the protoplasm of the botanist, and the contractile substance and sarcode of the zoologist" to be, "if not identical, yet in a high degree analogous substances."

Remak first extended the use of the term protoplasm from the layer which bore that name in the vegetable cell to the analogous element in the animal cell; but "it was Max Schultze, in particular, who by applying the name to the intracellular matrix, or contained matter, when divested of membrane, and by identifying this substance itself with sarcode, first fairly established protoplasm, name and thing, in its present position."

In England, however, it is Professor Huxley who, by his brilliant and well-known Essay on this subject in the *Fortnightly Review* for February, 1869, has acquired a prominence, though by no means a pre-eminence, all his own. Taking for his theme the "Physical Basis of Life," and treading in the track of that "host of investigators", of whom he tells us that they "have accumulated evidence, morphological, physiological, and chemical," in favour of that "immense unité de composition élémentaire dans tous les corps vivants de la nature," of which Payen wrote so clearly nearly thirty-five years ago; he combats "the widely-spread conception of life as a something which works through matter, but is independent of it"; and affirms, on the contrary, "that matter and life are inseparably connected, and that there is one kind of matter which is common to all living beings."

Notwithstanding the wide diversity that presents itself to our view in the countless varieties of living beings, it yet is true that all vegetable and animal tissues without exception, from that of the brightly coloured lichen on the rock, to that of the painter who admires or of the botanist who dissects it, are essentially one in composition and in structure. The microscopic fungi clustering by millions within the body of a single fly, the giant pine of California towering to the height of a cathedral spire, the Indian fig-tree covering acres with its profound shadow, animalcules minute enough to dance in myriads on the point of a needle, and the huge leviathan of the deep, the flower that a girl wears in her hair, and the blood that courses through her veins, are, each and all, smaller or larger multiples or aggregates of one and the same structural unit, and all therefore ultimately resolvable into the same identical elements. That unit is a corpuscle composed of oxygen, hydrogen, nitrogen, and carbon. Hydrogen, with oxygen, forms water; carbon, with oxygen, carbonic acid; and hydrogen, with nitrogen, ammonia, These three compounds—water, carbonic acid, and ammonia,—in like manner, when combined form protoplasm.

In all this, however, there is nothing new but the nomenclature.[1] But the case is widely altered when Mr. Huxley proceeds to assert that amid all the diversities of living things and living beings there exists a threefold unity: a unity of faculty, a unity of form, and a unity of substance. In relation to the first of these, for example, faculty, power, activity; according to Mr. Huxley, even *human* activities must be referred to three categories—contractility, alimentation, and reproduction; and for the lower forms of life, whether animal or vegetable, there are no fewer than these same three. The granulated, semi-fluid layer which constitutes the lining of the woody case of the nettle-sting is possessed of contractility. And in this possession of contractile substance, other plants are as the nettle, and all animals are as plants. Protoplasm is common to the whole of them; and this lining in the sting of the nettle is protoplasm. So that between the powers of the lowest plant or animal and those of the highest, the difference is one not of kind, but only of degree. The colourless blood-corpuscles in

---

[1] And this nomenclature, though new, is by no means improved. It is inexact, indefinite, indiscriminate, and therefore necessarily misleading. See below; especially pages 132, 135–142.

man and the other animals are identical with the protoplasm of the nettle; and he, not less than they, at first consisted of nothing more than an aggregation of such corpuscles. Protoplasm is their common constituent; in protoplasm they have their common origin. At last, as at first, all that lives, and every part of all that lives, is but—nucleated or unnucleated, modified or unmodified—protoplasm.

This series of assertions culminates in a dogma still more astounding. Protoplasm, from being "the basis," becomes "the matter of life." Apart from this matter, life is unknown. The "phenomena of life," however vast and varied, exhibit neither force nor faculty that is not derived from the chemical constituents of its material "basis." All the activities of life,— vegetable, animal, human; physical, intellectual, religious—arise solely (we are told) from "the arrangement of the molecules of ordinary matter." What reason is there, for instance, why thought should not be termed a property of thinking protoplasm, just as congelation is a property of water, or centrifugience of gas? Professor Huxley protests that he is aware of no reason. We call, he says, the several phenomena which are peculiar to water "the properties of water, and do not hesitate to believe that in some way or other they result from the properties of the component elements of water. We do not assume that something called *aquosity* entered into and took possession of the oxide of hydrogen as soon as it was formed, and then guided the aqueous particles to their places in the facets of the crystal or among the leaflets of the hoar-frost." Why, then, "when carbonic acid, water, and ammonia disappear, and in their place, under the influence of preexisting protoplasm, an equivalent weight of the matter of life makes its appearance," should we assume, in the living matter, the existence of "a something which has no representative or correlative in the unliving matter that gave rise to it"? Why imagine that into the newly formed hydro-nitrogenised oxide of carbon a something called vitality entered and took possession? "What better philosophic status has vitality than aquosity?"

These questions, as will presently appear, present no difficulty. They admit of answers too complete to leave room for further question. The only difficulty is that which presents itself when we attempt to determine Professor Huxley's relation to them. For incredible as it must seem to those not acquainted with the facts, the propositions above cited are at once

the subject of his affirmation and of his denial.
Dr. Stirling concludes his refutation of them in
a sentence to which Professor Huxley has attempted a reply. The sentence is this:—

"In short, the whole position of Mr. Huxley, that all organisms consist alike of the same life-matter, which life-matter is, for its part, due only to chemistry, must be 'pronounced untenable,'—nor less untenable the materialism he would found on it."[1]

And this is the reply:—

"The paragraph contains three distinct assertions concerning my views, and just the same number of utter misrepresentations of them." The first [that "all organisms consist alike of the same life-matter"] "turns on the ambiguity of the word 'same'"; the second [that this "life-matter is due only to chemistry"] "is in my judgment absurd, and certainly I have never said anything resembling it; while as to Number 3, one great object of my Essay[2] was to show that what is called 'materialism' has no sound philosophical basis."[3]

---

[1] "As Regards Protoplasm." By James Hutchinson Stirling, F.R.C.S., and LL.D. Edinburgh. Longmans, 1872, p. 58.

[2] "'One great object of my Essay,' says Mr. Huxley! Yes, truly; but what of the *other*—great, greater, and greatest—object? 'Utter misrepresentation!' The only utter misrepresentation concerned here is—— Pshaw! the whole thing is beneath speech." ("As Regards Protoplasm," *ut sup.*, p. 59.)

[3] "Yeast," in "Critiques and Addresses." Macmillan, 1873, p. 90.

Scientific Sophisms.

In rejoinder, Dr. Stirling cites "Mr. Huxley's own phrases" to prove that the alleged ambiguity does not exist: "There is such *a* thing as *a* physical basis or matter of life;" . . . or "*the* physical basis or matter of life." There is "a single physical basis of life," and through its unity, "the whole living world" is pervaded by "a threefold unity"—"namely a unity of power or faculty, a unity of form, and a unity of substantial composition."

On the second point; that "life-matter" is "due only to chemistry," Dr. Stirling is "pleased to think that Mr. Huxley has now come to consider such an opinion 'absurd,'" but repeats that "he has always, and everywhere, for all that, described his 'life-matter as due to chemistry,'" and adds, "Here are a few examples:"—

"'If the properties of water may be properly said to result from the nature and disposition of its component molecules, I can find no intelligible ground for refusing to say that the properties of protoplasm result from the nature and disposition of its molecules.'

"Is it possible for words more definitely to convey the statement that the properties of water and protoplasm are precisely on the same level, and that as the former are of molecular (physical, chemical) origin, so are the latter?

Again, after having told us that protoplasm is carbonic acid, water, and ammonia, 'which certainly possess no properties but those of ordinary matter,' he proceeds to speak as follows:—

"'Carbon, hydrogen, oxygen, and nitrogen are all lifeless bodies. Of these, carbon and oxygen unite in certain proportions, and under certain conditions, to give rise to carbonic acid; hydrogen and oxygen produce water; nitrogen and hydrogen give rise to ammonia. These new compounds, like the elementary bodies of which they are composed, are lifeless.'

"So far then, surely, I am allowed to say that these new compounds are *due* to chemistry. Observe now what follows:—

"'But when they' (the compounds) 'are brought together, under certain conditions, they give rise to the still more complex body protoplasm, and this protoplasm exhibits the phenomena of life. I see no break in this series of steps in molecular complication, and I am unable to understand why the language which is applicable to any one term of the series may not be used to any of the others.'

"Here, evidently, I am *ordered* by Mr. Huxley himself, not to change my language, but to characterise these latter results as I characterised those former ones. If I spoke then of ammonia, etc., as due to chemistry, so must I now speak of protoplasm, life-matter, as due to chemistry—a statement which Mr. Huxley

Scientific Sophisms. 119

not only orders *me* to make, but makes *himself*. Very curious all this, then. When I do what he bids me do, when I say what he says—that if ammonia, etc., are due to chemistry, protoplasm is also due to chemistry—Mr. Huxley turns round and calls out that I am saying an 'absurdity,' which he, for his part, 'certainly never said!' But let me make just one other quotation :—

"'When hydrogen and oxygen are mixed in a certain proportion, and an electric spark is passed through them, they disappear, and a quantity of water equal in weight to the sum of their weights appears in their place.'

"Now, no one in his senses will dispute that this is a question of chemistry, and of nothing but chemistry; but it is Mr. Huxley himself who asks in immediate and direct reference here :—

"'Is the case in any way changed when carbonic acid, water, and ammonia disappear, and in their place, under the influence of pre-existing living protoplasm, an equivalent weight of the matter of life makes its appearance?'

"Surely Mr. Huxley has no object whatever here but to place before us the genesis of protoplasm, and surely also this genesis is a purely chemical one! The very 'influence of pre-existing living protoplasm,'—which *pre-existence* could not itself exist for the benefit of the *first* protoplasm that came into existence,—is asserted to be in precisely the same case with reference to the one process as that of the electric spark with reference to the other. And yet, in the teeth of such passages, Mr. Huxley feels himself at liberty to say now, 'Statement Number 2 is, in my judgment, absurd, and *certainly I have never said anything resembling it.*' It is a pity to see a man in the position of Mr. Huxley so strangely *forget* himself!"

On the third head—Mr. Huxley's "materialism"—Dr. Stirling's refutation is equally conclusive, but at the same time, much too elaborate to admit of quotation here. No summary could do it justice; it must be read in its entirety. In this place, however, it does not concern us. It lies outside the sphere of our investigation. We are not now inquiring what esoteric meaning may be attached by Mr. Huxley to the language he has chosen to employ; nor even are we inquiring whether that language is compatible with any such meaning whatever. Our inquiry is much more simple. It is limited to the question of fact. Is it certain, is it demonstrable, is it scientifically true that the facts of the case are as stated by Mr. Huxley? On this very question of "mate-

Scientific Sophisms. 121

rialism," for instance, Mr. Huxley asserts that "all vital action" is but "the result of the molecular forces" of the physical basis; and consequently, to use his own words when addressing his Edinburgh audience, "the thoughts to which I am now giving utterance, and your thoughts regarding them, are but the expression of molecular changes in that matter of life which is the source of our other vital phenomena." With these words in their recollection, few persons would be disposed to differ from Mr. Huxley when he says that "most undoubtedly the terms of his propositions are distinctly materialistic."

But are they true?

"I know of no form of negation sufficiently explicit, comprehensive, and emphatic in which to reply to this question." The doctrines of Scientific Materialism, as above stated, in Professor Huxley's own words, are "so utterly at variance with the most familiar facts of chemistry that it is marvellous they should have so long passed unchallenged."[1]

1. To enter into detail. It is in no sense true

[1] "Unchallenged, that is," adds Dr. Elam, " on purely chemical grounds. On other issues, both relevant and irrelevant, they have been often objected to."

that protoplasm "breaks up" (as Professor Huxley says it does)[1] into carbonic acid, water, and ammonia, any more than it is true that iron, when exposed to the action of oxygen, "breaks up" into oxide of iron. A compound body can break up only into its constituent parts; and these *are not* the constituent parts of protoplasm. "To convert protoplasm into these three compounds requires an amount of oxygen *nearly double the weight* of the original mass of protoplasm; speaking approximately, every 100 lbs. of protoplasm would require 170 lbs. of oxygen."

2. "Under certain conditions," says Professor Huxley,[2] whereas, in point of fact, under *no possible* "*conditions*" can carbonic acid, water, and ammonia, when brought together, "give rise to the still more complex body protoplasm." "Not even on paper can any multiple, or any combination whatever of these substances, be

[1] "The matter of life . . . breaks up . . . into carbonic acid, water, and ammonia, which certainly possess no properties but those of ordinary matter." (Professor Huxley, in *The Fortnightly Review*, February, 1869.)

[2] "But when they [the "lifeless compounds" carbonic acid, water, and ammonia] are brought together, under certain conditions, they give rise to the still more complex body protoplasm, and this protoplasm exhibits the phenomena of life." (*Ibid.*)

made to represent the composition of protoplasm; much less can it be effected in practice. Carbonic acid (CO_2), water (H_2O), and ammonia (NH_3), cannot by any combination be brought to represent $C_{36}H_{26}N_4O_{10}$, which is the equivalent of protein or protoplasm.

3. "But the most incredible of all the errors, if it be not simply a mystification, is found in the comparison between the formation of water from its elements and the origination of protoplasm. Hydrogen and oxygen doubtless unite to form an equivalent weight of water; that is, an amount of water equalling in weight the combined weights of the hydrogen and the oxygen; and Professor Huxley asks, 'Is the case in any way changed when carbonic acid, water, and ammonia disappear, and in their place, under the influence of pre-existing protoplasm, an equivalent weight of the matter of life makes its appearance?'

"The answer is, Certainly; the case is changed in every possible way in which a process, whether chemical or otherwise, can be changed. But it must also be premised that the fact as stated is *not true*, that when these three substances disappear, under certain conditions, an '*equivalent weight* of the matter of life makes its appearance.' Every chemist knows what an 'equivalent weight' means; knows also that there can be no weight of protoplasm 'equivalent,' chemically speaking, to any amount of carbonic acid, water, and ammonia, that may or can have disappeared. These are simple, well-known, and understood chemical facts, and need no discussion.

4. "But granting for the moment, and for the sake of argument, that these bodies disappear, and that protoplasm appears, it is manifest—almost too manifest to require stating—that there is *no resemblance whatever* in the two processes by which the results which Professor Huxley considers identical are obtained. In the formation of water, the whole of its constituent parts combine to form an equal weight of the compound; the case is entirely otherwise with regard to protoplasm, for here the so-called elements *do not combine at all.* On the contrary, they are uncombined or decomposed, by a process and by affinities most assuredly unknown in our laboratories. The carbonic acid and the ammonia are certainly decomposed, and whilst the carbon and nitrogen are assimilated, and add to the bulk of the plant, part of the oxygen is eliminated by the leaves, and part is destined to the performance of various functions in the economy."

And yet it is in this complex programme of decomposition, selection, fixation, and rejection, that we are asked to see nothing more than a process analogous to the formation of water from its elements; and Professor Huxley can see "no break." How wide must a chasm be before it is visible to an Evolutionist?

5. "Under certain conditions" only, and not otherwise, do the "lifeless compounds" aforesaid "give rise to the still more complex body protoplasm, and this protoplasm exhibits the phenomena of life." What are these conditions? The answer is that "when carbonic acid, water, and ammonia disappear, and in their place," "an equivalent weight of the matter of life makes its appearance," this appearance and disappearance are due to "the influence of pre-existing protoplasm."

From this it has been hastily, but most unwarrantably, assumed that vitality is a result of some particular arrangement of the molecular particles, the chemical constituents of protoplasm. In other words, that life is a product of protoplasm. But this proposition is demonstrably untrue.

Protoplasm, as known to us, is non-existent except as produced " under the influence of preexisting protoplasm." Water, ammonia, and carbonic acid cannot combine to form protoplasm unless a principle of life preside over the operation. Unless under those auspices, the combination never takes place. At present, whenever assuming its presidential functions, this principle of life appears invariably to be embodied in pre-existing protoplasm; but no one denies that there was a time when the fact was otherwise. Time was—as geology leaves no room for doubt—when our globe consisted wholly of inorganic matter, and possessed not one single vegetable or animal inhabitant. In that time it was not only possible for life, without being previously embodied, to mould and vivify inert matter, but the possible was the actual too. For if matter, inorganic and inanimate, had not been organized and animated by unembodied life, it would have remained inorganic and inanimate to this day. Those who would escape this conclusion have only one possible alternative. They must suppose that death gave birth to life. That matter, absolutely inert and lifeless, did spontaneously exert itself with all the marvellous energy indispensable for its conversion into living matter. That in making this exertion it wielded powers of which it was not possessed; powers which, under the conditions of the case, it could not have acquired, except by exercising them before it had acquired them. That, absolutely inert as it was, it yet made this impossible exertion; and, lifeless as it was, it created life.

To reject incredible absurdities like these is to admit that originally protoplasm must have been produced by life not previously embodied; but to admit this and yet to suppose that when, as now, embodied life is observed to give birth to new embodiments, the operative force belongs not to the life itself, but to its protoplasmic embodiment, is " much the same as to suppose that when a tailor, dressed in clothes of his own making, makes a second suit of clothes, this latter is the product not of the tailor himself, but of the clothes he is wearing."[1] Life therefore is not a product of protoplasm.

d. Nor is it a property of protoplasm.

By the property of an object is meant, in scientific speech, not merely something belonging to the object, but also that it is a thing without which the object could not subsist. Thus, fluidity, solidity, and vaporisation are "properties" of water, because matter which did not liquefy, congeal, and evaporate at

[1] "Old-fashioned Ethics, and Common-sense Metaphysics." By William Thomas Thornton. Macmillan, 1873, chap. iv. p. 167. ("Huxleyism.")

different temperatures would not be water. It is the exhibition of these phenomena, in conjunction with certain others, that constitutes the "aquosity" or wateriness of water. But in no such sense, nor in any sense whatever, is life or "vitality" essential to that species of matter which Mr. Huxley calls "matter of life," or protoplasm. Take from water its aquosity, and water ceases to be water; but you may take away vitality from protoplasm, and yet, according to Mr. Huxley's own affirmation,[1] leave protoplasm as much protoplasm as before. Whatever therefore may be the relation which vitality bears to protoplasm, it is a relation totally different from that which aquosity bears to water. When therefore Professor Huxley asks: "What better philosophic status has vitality than aquosity?" we answer:—Protoplasm can do perfectly well without "vitality;" but water cannot for a moment dispense with "aquosity." "Protoplasm, whether living or lifeless, is equally itself; but unaqueous water is unmitigated gibberish."[2] Since then, as Mr. Huxley affirms, protoplasm even when

[1] "*Living or dead*," says Mr. Huxley: "If the phenomena exhibited by water are its properties, so are those presented by protoplasm, living or dead, its properties."
[2] Thornton's "Old-fashioned Ethics," *ut sup.*, p. 165.

deprived of its vitality is still protoplasm, it is axiomatically evident that vitality is not indispensable to protoplasm, and is therefore not a "property" of protoplasm.

7. But this question of Mr. Huxley's is further noticeable on account of the connection in which it is found; a connection highly significant in relation to its author's disclaimer of "materialism." In varying phrase, but always to the same effect, in three short consecutive sentences he thrice reiterates the question :—

"What justification is there then for the assumption of the existence in the living matter of a something which has no representative or correlative in the not-living matter that gave rise to it? What better philosophic status has vitality than aquosity? And why should vitality hope for a better fate than the other *itys* which have disappeared since Martinus Scriblerus accounted for the operation of the meat-jack by its inherent meat-roasting quality, and scorned the materialism of those who explained the turning of the spit by a certain mechanism worked by the draught of the chimney?"[1]

"This," replies Dr. Elam, "is very amusing —no one can be more so than Professor Huxley; —a little perception of facts and analogies would make it perfect. The answer is obvious, if answer is required. All these are machines

[1] *Fortnightly Review*, February, 1869, p. 140.

which man has made, and can again make by the use of well-known forces and material which he can combine at will; it is not therefore necessary to hypothecate any other force or principle. When man can make any, even the simplest organism, out of inorganic matter, then shall we be compelled to acknowledge that chemical and other forces are sufficient, and that the hypothesis of a vital principle has had its day and may cease to be. To Professor Huxley's illustration I will respond seriously when he has demonstrated to me that meat-jacks have been developed from the beginning of time only and exclusively under the immediate contact and influence of pre-existing meat-jacks. Until then the analogy is scarcely close enough to need refutation or discussion."[1]

8. Mr. Huxley, as above cited, refuses to recognise the distinction between dead protoplasm and that which lives. Other authorities however, and especially the Germans who have led the way in this investigation, say expressly that whether the same elements are to be referred to the protoplasmic cells equally after death as before it is a matter entirely unknown. While this is so it is evident that Mr. Huxley's

[1] *Contemporary Review*, September, 1876, p. 558 *et seq.*

chemical analysis of dead protoplasm cannot be regarded as decisive for that which is not dead. And yet, throughout his whole argument, he builds on this same chemical analysis as if it were decisive. Thus he speaks of mutton as "once the living protoplasm," now the "same matter altered by death" and cookery, but yet as not being by these alterations rendered "incompetent to resume its old functions as matter of life."[1] He speaks of its being subjected to "*subtle influences*" which "will convert the dead protoplasm into the living protoplasm"—which will "raise the complex substance of dead protoplasm to the higher power, as one may say, of living protoplasm."[2] In all this, as throughout, when he speaks of dead matter of life and living matter of life, not only is there no hint of any difference in chemical constitution, or in "arrangement of molecules," between the dead and the living, but when, in anticipation of such difference, he alludes to it at all, it is only to pronounce it "frivolous."[3]

So be it. Let the identity of protoplasm, "living or dead," as assumed by Mr. Huxley,

[1] *Fortnightly Review*, February, 1869, p. 137.
[2] *Ibid.*, p. 138.
[3] *Ibid.*, p. 135.

be—at least for the moment, and for the sake of the argument—conceded. What then? The properties of protoplasm, as we have seen, are altogether dependent upon the arrangement of its constituent atoms. But protoplasm in one of these conditions (*i.e.*, dead) manifests *passive* properties only; while, in the opposite condition,—without any change, *i.e.*, any known or knowable change, in its chemical properties or molecular arrangement,—we find it exercising a vast variety of *active* properties, assimilation, contraction, reproduction; not to mention thought, feeling, and will. Here then we have an effect, or rather a whole train of effects most marvellous,—*without a cause*, a conclusion that the most enthusiastic Evolutionist would hesitate to pronounce in "general harmony with scientific thought."[1] From this impossible, and yet inevitable conclusion there is no possible escape except (1) by hypothecating a change, mechanical or chemical, of which, by Professor Huxley's own confession, we can have no possible knowledge,[2] and on which therefore "we have no right to speculate;"

---

[1] "Belfast, Address," *ut sup.*, p. 58: "The strength of the doctrine of evolution consists . . . in its general harmony with scientific thought."

[2] *Fortnightly Review.*

or (2) by confessing that the "subtle influences" invoked by Mr. Huxley to eke out the deficiencies of protoplasmic chemistry are nothing else than—under another name—that very same vital force or vital principle in which it is now so unfashionable and so unscientific to believe.[1]

9. In truth, however, the fulcrum on which Mr. Huxley's protoplasmic materialism rests is a single inference from a chemical analogy. But analogy, which is never identity, though often mistaken for it, is apt to betray. The difference which it covers may be essential, while the likeness it reveals may be inessential —as far as the conclusion is concerned. The analogy to which Mr. Huxley trusts has two references: one to chemical composition, and one to a certain stimulus that determines it. In both of these the analogy fails: in both it can only seem to succeed by discounting the elements of difference that still subsist.

It cannot be denied that protoplasm is a chemical substance. It cannot be denied that protoplasm is a physical substance. Both physically and chemically, water (as a compound of hydrogen, and oxygen) and protoplasm (as

[1] Dr. Elam's "Automatism and Evolution" (*ut sup.*), p. 560.

a compound of carbon, hydrogen, oxygen, and nitrogen) are clearly analogous. So far as it is on chemical and physical structure that the possession of distinctive properties in any case depends, both bodies may be said to be on a par. So far the analogy must be allowed to hold ; so far, but no farther. "One step farther, and we see not only that protoplasm has, like water, a chemical and physical structure ; but that, unlike water, it has also an organised or organic structure. Now this, on the part of protoplasm, is a possession in excess ; and with relation to that excess there can be no grounds for analogy." When therefore Mr. Huxley says, "If the phenomena exhibited by water are its properties, so are those presented by protoplasm, living or dead, its properties," the answer is, " Living or dead?" organic or inorganic? That alternative is simply slipped in and passed ; but it is in that alternative that the whole matter lies. Chemically, dead protoplasm is to Mr. Huxley quite as good as living protoplasm. It is this dead protoplasm which he finds so delectable in the shape of bread, lobster, mutton. But then it is to be remembered that it is only these—as being inorganic—that can be placed on the same level as water; while

Scientific Sophisms. 135

living protoplasm is not only unlike water, but it is unlike dead protoplasm. Living and dead protoplasm are identical only as far as chemistry is concerned (if indeed so far as that); it is therefore evident, consequently, that difference between the two cannot depend on that in which they are identical; *i.e.*, cannot depend on the chemistry.

Life, then, is something else than the result of chemical or physical structure, and it is in another sphere than those of physics or chemistry that its explanation must be found. It is thus that, lifted high enough, the light of the analogy between water and protoplasm is seen to go out. Water, like its constituent elements, has only chemical and physical qualities; like them, it is still inorganic. But not so in protoplasm, where, together with retention of the chemical and physical likeness, there is the addition of the unlikeness of life, of organization, and of ideas. But this addition is a world in itself: a new and higher world, the world of a self-realizing thought, the world of an *entelechy*. The relation of the organic to the inorganic— of protoplasm dead to protoplasm alive—is not an analogy, but an antithesis: The antithesis of antitheses. In it, in fact, we are in presence of the one impassable gulf—"that gulf which Mr. Huxley's protoplasm is as powerless to efface as any other material expedient that has ever been suggested since the eyes of men first looked into it—the mighty gulf between death and life."[1]

10. "Protoplasm is the clay of the potter, which, bake it and paint it as he will, remains clay, separated by artifice, and not by nature, from the commonest brick or sun-dried clod." On this it has been justly observed that " Mr. Huxley puts emphatically his whole soul into this sentence, and evidently believes it to be, if we may use the word, a *clincher*." But the answer is easy. The assertion that all bricks, being made of clay, are the same thing, is one that involves its own limitation. Yes, undoubtedly, we answer, if they are made of the same clay. The bricks are identical if the clay is identical; but, on the other hand, by as much as the clay differs will the bricks differ. And, similarly, all organisms can be identified only if their composing protoplasm can be identified. But when, from indefinite generalizations, we descend to definite particulars, this identification is found to be impossible.

Mr. Huxley's entire theory may be summed

[1] Dr. Stirling: "As Regards Protoplasm," p. 41.

up in two propositions :—First, "That all animal and vegetable organisms are essentially alike in power, in form, and in substance;" Second, "That all vital and even intellectual functions are the properties of the molecular disposition and changes of the material basis (protoplasm) of which the various animals and vegetables consist." In both propositions the agent of proof is this same alleged material basis of life, or protoplasm. To establish the first, Mr. Huxley endeavours to identify all organisms (animal and vegetable) in protoplasm. To establish the second, by means of inference from a simple chemical analogy he assigns vitality, and even intellect, to the molecular constituents of the protoplasm, in connection with which they are exhibited.

The second of these propositions has already been examined and refuted. It has been shown[1] that life is not a property of protoplasm; that it is not a product of protoplasm; and that vitality and protoplasm are not inseparable. Be protoplasm what it may, vital and intellectual functions are not the products of its molecular constitution.

It is the first of these two propositions which now remains to be examined. Is protoplasm,

[1] In paragraphs 5, 6, 8, and 9, pp. 120-129.

as alleged by Mr. Huxley, an actual life-matter, everywhere identical in itself, and one which consequently everywhere involves the identity of all the various organs and organisms which it is assumed to compose? The bricks, says Mr. Huxley, are the same because the clay is the same. But is the clay the same? Can it be identified, as Mr. Huxley alleges, by a three-fold unity of faculty, of form, of substance?

To begin then with this simplest question, that of substance. Are all samples of protoplasm identical, first, in their chemical composition, and, second, under the action of the various re-agents? This cannot be affirmed. And it is against the affirmation of this that "we point to the fact of much chemical difference obtaining among the tissues, not only in the *proportions* of their fundamental elements, but also in the *addition* (and proportion as well) of such others as chlorine, sulphur, phosphorus, potass, soda, lime, magnesia, iron, etc. Vast differences vitally must be legitimately assumed for tissues that are so different chemically."[1]

As to the alleged unities of form and power in protoplasm, according to Stricker,[2] "Proto-

[1] Dr. Stirling: "As Regards Protoplasm," p. 29.
[2] Whom Professor Huxley calls, "My valued friend Professor Stricker." ("*Yeast*," in "Critiques and Addresses,"

plasm varies almost infinitely in consistence, in shape, in structure, and in function.

"In consistence, it is sometimes so fluid as to be capable of forming in drops; sometimes semi-fluid and gelatinous; sometimes of considerable resistance. In shape—for to Stricker the cells are now protoplasm—we have club-shaped protoplasm, globe-shaped protoplasm, cup-shaped protoplasm, bottle-shaped protoplasm, spindle-shaped protoplasm, branched, threaded, ciliated protoplasm, circle-headed protoplasm, flat, conical, cylindrical, longitudinal, prismatic, polyhedral, and palisade-like protoplasm. In structure, again, it is sometimes uniform and sometimes reticulated into interspaces that contain fluid.

"In function, lastly, some protoplasm is vagrant, and of unknown use. Some again produces pepsine, and some fat. Some at least contain pigment. Then there is nerve-protoplasm, brain-protoplasm, bone-protoplasm, muscle-protoplasm, and protoplasm of all the other tissues, no one of which but produces its own kind, and is uninterchangeable with the rest. Lastly, on this head, we have to point to the overwhelming fact that there is the infinitely different protoplasm of the various infinitely different plants and animals, in each of which its own protoplasm, as in the case of the various tissues, but produces its own kind, and is uninterchangeable with that of the rest."[1]

The evidence in refutation of Mr. Huxley's first proposition is thus seen to be overwhelm-

p. 89.) Stricker: with whom, says Dr. Stirling, "for the production of his 'Handbuch,' there is associated every great histological name in Germany." (Pref., p. 3.)

[1] "As Regards Protoplasm," pp. 30, 31.

ing. In view of the nature of microscopic science; in view of the results hitherto obtained as regards nucleus, membrane, and entire cell, even in view of the supporters of protoplasm itself; Mr. Huxley's assertion of a physical matter of life is untenable.[1] But even if that "matter of life" were granted, the reasons innumerable, and even irrefragable, would still remain to compel us—as now they do actually compel us—to acknowledge in it, not indeed the "identity" now claimed, but rather "an infinite diversity" in power, in form, and in substance. No wonder that the bricks are not the same: with this "infinite diversity" in the clay.

11. Nor is this fundamental diversity in any way altered or diminished by the convertibility of which Mr. Huxley speaks. On the contrary, that convertibility, as alleged by Mr. Huxley,

---

[1] The position here maintained—in opposition to Mr. Huxley—is supported by an important dictum of Professor Tyndall:—" When the contents of a cell are described as perfectly homogeneous, as absolutely structureless, because the microscope fails to distinguish any structure, then I think the microscope begins to play a mischievous part. A little consideration will make it plain to all of you that the microscope can have no voice in the real question of germ structure."—*Fragments of Science:* First Edition, p. 155.

establishes. the antecedent diversity. If the diversity were non-existent, there would be no room for the alleged process of convertibility. And yet, as used by him, this same convertibility is employed to stamp protoplasm (and with it life and intellect) into an indifferent identity. In order that there may be "no break" between the lowest functions and the highest—between the functions of the fungus and the functions of man—he has "endeavoured to prove," he tells us, that the protoplasm of the lowest organisms is "essentially identical with, and most readily *converted* into that of any animal."[1] And on this alleged reciprocal *convertibility* of protoplasm he founds an inference of identity, as well as of the further conclusion that the functions of the highest, not less than of the lowest animals, are but the molecular manifestations of the protoplasm which is common to all.

"Is this alleged reciprocal *convertibility* true, then? Is it true that every organism can digest every other organism, and that thus a relation of identity is established between that which digests and whatever is digested?

"These questions place Mr. Huxley's general enterprise, perhaps, in the most glaring light yet; for it is very evident

---

[1] "Lay Sermons," p. 138.

that there is an end of the argument if all foods and all feeders are essentially identical both with themselves and with each other. The facts of the case, however, I believe to be too well known to require a single word here on my part. It is not long since Mr. Huxley himself pointed out the great difference between the foods of plants and the foods of animals; and the reader may be safely left to think for himself of *ruminantia* and *carnivora*, of soft bills and hard bills, of molluscs and men. Mr. Huxley talks feelingly of the possibility of himself feeding the lobster quite as much as of the lobster feeding him; but such pathos is not always applicable: it is not likely that a sponge would be to the stomach of Mr. Huxley any more than Mr. Huxley would be to the stomach of a sponge.

"But a more important point is this, that the functions themselves remain quite apart from the alleged convertibility. We can neither acquire the functions of what we eat, nor impart our functions to what eats us. We shall not come to fly by feeding on vultures, nor they to speak by feeding on us. No possible manure of human brains will enable a corn-field to reason. But if functions are inconvertible, the convertibility of protoplasm is idle. In this inconvertibility, indeed, functions will be seen to be independent of mere chemical composition. And that is the truth: for function there is more required than either chemistry or physics."[1]

[1] Dr. Stirling: "As Regards Protoplasm," p. 50.

12. As of the bricks, then, so of the clay: it is not identical, and it is not convertible. But Evolution dies hard, and Mr. Huxley in the last resort falls back upon protoplasm "variously *modified.*" But where are we to begin, *not* to have modified protoplasm? Mr. Huxley begins with the sting of the nettle, but even there the protoplasm is already modified; and we have the authority of Rindfleisch for asserting that "in every different tissue we must look for *a different initial term* of the productive series."

Besides: there are in protoplasm generic or specific differences; differences not merely of degree, but of kind. Some of these are indicated by Mr. Huxley himself, when he tells us that plants alone are capable of assimilating inorganic matter; while animals assimilate organic matter only. Others must be admitted "for the overwhelming reason that an infinitude of various kinds exist in it, each of which is self-productive and uninterchangeable with the rest." Brain-protoplasm is not bone-protoplasm, nor the protoplasm of the fungus the protoplasm of man. "If the cornea of the eye and the enamel of the teeth are alike but modified protoplasm, we must be pardoned for thinking more of the adjective than of the substantive. Our wonder is how, for one idea, protoplasm could become one thing here, and, for another idea, another so different thing there. We are more curious about the modification than the protoplasm. In the difference, rather than in the identity, it is indeed that the wonder lies.

"Here are several thousand pieces of protoplasm; analysis can detect no difference in them. They are to us, let us say, as they are to Mr. Huxley, identical in power, in form, and in substance; and yet on all these several thousand little bits of apparently indistinguishable matter an element of difference so pervading and so persistent has been impressed, that of them all, not one is interchangeable with another! Each seed feeds its own kind. The protoplasm of the gnat will no more grow into the fly than it will grow into an elephant. Protoplasm is protoplasm; yes, but man's protoplasm is man's protoplasm, and the mushroom's the mushroom's."[1] The difference is one of kind, not of degree; and that difference the word "modification," though it may indeed sometimes conceal, will never be able to efface.

13. In closing this brief review of Mr. Huxley's

[1] "As Regards Protoplasm," p. 58.

Scientific Sophisms. 145

doctrine, it will be found not unimportant to notice some particulars which characterise Mr. Huxley's own position in relation to it. Foremost among these is the nomenclature which Mr. Huxley has chosen to employ.

The protoplasmic pellicle, "the formative protoplasmic layer" in vegetable cells, was regarded by Von Mohl as a structure of special importance, distinct from the cell-contents, and was named by him, in 1844, "the primordial utricle." This primordial utricle has since been called protoplasm by Professor Huxley, although some years previously he had restricted the term protoplasm to the matter *within the primordial utricle*, which matter he at that time regarded as nothing more than an "accidental anatomical modification" of the endoplast, and of little importance.[1] "The nucleus, and with it the protoplasm, Mr. Huxley thought, *exerted no peculiar office*, and *possessed no metabolic power*. But Mr. Huxley has changed his views without one word of explanation concerning the facts which led him to modify them, or even an acknowledgment that he had changed them. Mr. Huxley now considers 'protoplasm' of the first importance. . . . His 'endoplast' and 'peri-

[1] "The Cell Theory:" *Medical Chirurgical Review*, October, 1853.

plastic substance', of 1853 together constitute his 'protoplasm' of 1869."[1]

14. "In order to convince people that the actions of living beings are not due to any mysterious vitality or vital force or power, but are in fact physical and chemical in their nature, Professor Huxley gives to matter which is *alive*, to matter which is *dead*, and to matter which is *completely changed by the process of roasting or boiling*, the very same name. 'Mutton contained protoplasm of the same nature as was found in every living thing.' 'As he spoke, he was wasting his stock of protoplasm, but he had the power of making it up again by drawing upon the protoplasm of some other animal— say a sheep. (Laughter.)' The matter of sheep and mutton and man and lobster and egg is the *same*, and, according to Huxley, one may be *transubstantiated* into the other. But how? By 'subtle influences,' and 'under sundry circumstances,' answers this authority. And all these things *alive*, or *dead*, or *dead* and *roasted*, he tells us are made of protoplasm, and he affirms this protoplasm is the *physical basis* of

[1] "Protoplasm; or Matter and Life." By Lionel S. Beale, M.B., F.R.S. Third Edition. London: Churchill, 1874, pp. 90, 91.

life, or the basis of *physical life*.[1] But is it not hard that the discoverer of '*subtle influences*' should laugh at the fiction of '*vitality*'! By calling things which differ from one another in many qualities by the same name, Huxley seems to think he can annihilate distinctions, enforce identity, and sweep away the difficulties which have impeded the progress of previous philosophers in their search after unity. Plants and worms and men are all protoplasm, and protoplasm is albuminous matter, and albuminous matter consists of four elements, and these four elements possess certain properties, by which properties all differences between plants and worms and men are to be accounted for. Although Huxley would probably admit that a worm was not a man, he would tell us that by 'subtle influences' and 'under sundry circumstances,' the one thing might be easily converted into the other, and not by such nonsensical fictions as 'vitality,' which can neither be weighed, measured, nor conceived. But, in

[1] [Note by Dr. Beale :] The heading of his lecture as published in *The Scotsman* for November 9, 1868, is "The Bases of Physical Life," while his communication in *The Fortnightly*, February 1, 1869, referred to by him as this same lecture, is entitled "The Physical Basis of Life." The iron basis of the candle, and the basis of the iron candle, are expressions evidently interchangeable.

science, it is not fair to indulge in word-tricks and equivocal illustrations, nor is it justifiable to make use of misleading similes."[1]

15. "I think Professor Huxley is the first observer who has spoken of the cell in its entirety as a mass of protoplasm, and the only one who has ever asserted that any tissue in nature is composed throughout of matter which can properly be regarded as one in kind. This view is quite irreconcilable with many facts, some of which have been alluded to by Mr. Huxley himself. I doubt if in the whole range of modern science it would be possible to find an assertion more at variance with facts familiar to physiologists than the statement that 'beast and fowl, reptile and fish, mollusc, worm, and polype,' are composed of 'masses of protoplasm with a nucleus,' unless it be that still more extravagant assertion that what is ordinarily termed a cell or elementary part is a *mass of protoplasm;* for can anything be more unlike the semi-fluid, active, moving matter of amœba protoplasm, than the hard, dry, passive, external part of a cuticular cell or of an elementary part of bone?"[2]

[1] Dr. Beale's "Protoplasm," *ut sup.*, pp. 95, 96.
[2] *Ibid.*, pp. 97, 98.

Scientific Sophisms.

"Huxley makes no difference between dead and living and roasted matter, and he confuses together the living thing, the stuff upon which it feeds, and the things formed by it, or which result from its death. A muscle is protoplasm; nerve is protoplasm; a limb is protoplasm; the whole body is protoplasm, and of course bone, hair, shell, etc., are as much 'the physical basis of life' as albuminous matter and roast mutton. But surely it would be less incorrect to speak of such 'protoplasms' as the physical basis of *death* or the physical basis of *roast* than to call dead and roasted matter the physical basis of *life*. . . . Huxley says lobster-protoplasm may be converted into human protoplasm, and the latter again turned into living lobster. But the statement is incorrect, because in the process of assimilation what was once 'protoplasm' is entirely disintegrated, and is not converted into the new tissue in the form of protoplasm at all; and I must remark that sheep cannot be transubstantiated into man, even by 'subtle influences,' nor can dead protoplasm be converted into living protoplasm, or a dead sheep into a living man. And what is gained by calling the matter of dead roast mutton and that of a living growing sheep by the same name? If the last is the physical basis of *life*, one does not see how the first can be so too, unless roast mutton and living sheep are identical."[1]

Plain-speaking, this of Dr. Beale's; but its irresistible force is found in the well-earned celebrity of its author—"the foremost microscopist of the English-speaking world."[2]

[1] Dr. Beale's "Protoplasm," *ut sup.*, pp. 100, 101.
[2] "Beale's protoplasmic theory now takes the place of the cell theory. General opinion is now in accord, as respects the facts, with Dr. Beale's statements on the

16. "It is significant that Huxley himself, some sixteen years ago, drew a distinction between living and non-living matter, which he now, without any explanation, utterly ignores. He remarked that the stone, the gas, the crystal, had an *inertia*, and tended to remain as they were unless some external influence affected them; but that living things were characterised by the very opposite tendencies. He referred also to 'the faculty of pursuing their own course' and the 'inherent law of change in living beings.' In 1853, the same authority actually found fault with those who attempted to reduce life to 'mere attractions and repulsions,' and 'considered physiology simply as a complex branch of mere physics.' He also remarked that 'vitality is a property inherent in *certain kinds* of matter.'"[1] Now, however, as we have seen, there is but one kind of matter, "variously modified;" and "vitality" has no better status than "aquosity!"

17. Nor is it less "significant" to note Mr. Huxley's various, though incidental admissions, and to contrast them with the dogmatism of his

nucleus in 1860." (Dr. John Drysdale: "Protoplasmic Theory of Life." London, 1874.)

[1] Dr. Beale: *ut sup.*, p. 101.

mere assertions. We look for certainty and find only probability : *e.g.*,—" It is more than probable that *when* the vegetable world *is* thoroughly explored we *shall* find all plants in possession of the same powers." The premises then have still to be collected ; and yet the conclusion has been confidently proclaimed. Compare this "more than probable" vaticination concerning vegetables with the positive assertion "that *the powers* of ALL *the different forms* of living things were substantially *one*, that their forms were substantially *one*, and, finally, that their composition was also substantially *one*."[1] Again, he says, " *So far* as the conditions of the manifestations of the phenomena of contractility have *yet* been studied." Now this "so far" is not "yet" by any means "very far." But what is meant by "the manifestations of the phenomena"? The manifestations *are* the phenomena! and they completely refute Mr. Huxley's latest theory. Again, we hear that it is "the rule *rather* than the exception," or that "weighty authorities have *suggested*" that such and such things " probably occur," or, while contemplating the nettle-sting, that such "*possible* complexity" in other cases "*dawns* upon one." On other occasions he admits that

[1] *Scotsman*, November 9, 1868.

"perhaps it would not yet be safe to say that *all* forms," etc. Nay, not only does he directly *say* that "it is by no means his intention to suggest that there is no difference between the lowest plant and the highest, or between plants and animals," but he directly proves what he says, for he demonstrates in plants and animals an *essential difference of power*. Plants *can* assimilate inorganic matters, animals can *not*, etc.

18. "Mr. Huxley's ideas as to the composition of protoplasm have already been noticed, and it has been shown that they are clearly opposed to the known facts of science. Here a simple alternative presents itself; either Mr. Huxley is familiar with the elementary facts of organic chemistry, in which case he would be aware of the impossibility of such a composition; or he is not so, on which supposition it was at least indiscreet to found an important practical doctrine like that of human automatism on a purely fanciful chemical theory. Which alternative is to be adopted may perhaps receive some illustration from a parallel passage in the essay 'On the Formation of Coal,'[1]

[1] "Critiques and Addresses," pp. 109, 110.

Scientific Sophisms. 153

where, referring to the burning of coal, it is said:—

"'Heat comes out of it, light comes out of it, and if we could gather together all that goes up the chimney, and all that remains in the grate of a thoroughly-burnt coal-fire, we should find ourselves in possession of a quantity of carbonic acid, water, ammonia, and mineral matters, *exactly equal* in weight to the coal!'

"It requires but the most elementary acquaintance with the subject to recognise that the 'quantity' of these products would be at least twice, probably thrice, as great as the original weight of the coal. A due consideration and comparison of these facts will enable the reader to estimate at its true value the *science* from which such stupendous consequences are so confidently deduced."[1]

19. "How such doctrines came to be received can only be accounted for in Professor Huxley's own words when treating on some other antagonistic 'teaching,' which he says was only 'tolerable on account of the ignorance of those by whom it was accepted.' Referring to some anatomical question, he says further that 'it would, in fact, be unworthy of serious refutation,

[1] Dr. Elam: "Automatism and Evolution;" *Contemporary Review*, October, 1876, pp. 729, 730.

except for the general and natural belief that deliberate and reiterated assertions must have some foundation.'[1] It is by this time tolerably clear that Professor Huxley's 'Chemistry of Life' has no foundation except that of 'deliberate and reiterated assertion.'"[2]

But "if such be the case with the chemistry, what is to be said for the argument founded upon it, or attached to it—if, indeed, argument it can be called?" It has now been tried, and found wanting, in every particular. It is condemned by its own admissions. It is condemned by the magnitude of its assumptions. It is condemned by its antagonism to notorious, facts, and its violation of established principles. And the sentence which has followed condemnation is not less just than severe :—

"I cannot more appropriately conclude this notice of the doctrine of 'The Physical Basis of Life,' than with an extract from the author's own anthology of criticism, where,[3] speaking of the theory of creation, he says :—

"'That such verbal hocus-pocus should be received as

---

[1] "Evidence as to Man's Place in Nature," p. 85.
[2] Dr. Elam: *Contemporary Review*, September, 1876, p. 555.
[3] Professor Huxley's "Lay Sermons," p. 285.

science will one day be regarded as evidence of the low state of intelligence in the nineteenth century, just as we amuse ourselves with the phraseology about nature's abhorrence of a vacuum, wherewith Torricelli's compatriots were satisfied to explain the rise of water in a pump.'"[1]

---

[1] Dr. Elam: *Contemporary Review*, October, 1876 : p. 732.

CHAPTER VIII.

THE THREE BEGINNINGS.

"GIVE me matter," said Kant, "and I will explain the formation of a world; but give me matter only, and I cannot explain the formation of a caterpillar." This dictum is widely different from that of Professor Tyndall, who discerns in matter alone "the promise and potency of all terrestrial life." To the same effect is his eulogium on the Italian philosopher, Giordano Bruno, of whom he tells us[1] that "he came to the conclusion that Nature in her productions does not imitate the technic of man. Her process is one of unravelling and unfolding. The infinity of forms under which matter appears were not imposed upon it by an external artificer; by its own intrinsic force and virtue it brings these forms forth. Matter is not the mere naked, empty *capacity* which philosophers have pictured her to be, but the universal

[1] "Belfast Address," pp. 19, 20.

mother who brings forth all things as the fruit of her own womb."

In this opinion, Bruno and his eulogist are at one. In his controversy with Mr. Martineau, a year after the delivery of the Belfast Address, Dr. Tyndall credits "pure matter with the astonishing building power displayed in crystals and trees."[1] He "figures" to himself the embryological growth of the babe, and its " appearance in due time, a living miracle, with all its organs and all their implications." He dilates, justly and forcibly, on the wonders of eye and ear: the eye "with its lens, and its humours, and its miraculous retina behind;" the ear " with its tympanum, cochlea, and Corti's organ—an instrument of three thousand strings, built adjacent to the brain, and employed by it to sift, separate, and interpret, antecedent to all consciousness, the sonorous tremors of the external world. All this has been accomplished," the ells us, "not only without man's contrivance, but without his knowledge, the secret of his own organization having been withheld from him since his birth in the immeasurable past, until the other day." And then he adds, "Matter I define as that mysterious thing by

[1] "Materialism and its Opponents," p. 594.

which all this is accomplished."[1] No wonder then that Bruno should be lauded for his "closer" approximation "to our present line of thought." [2]

But this expression—"our present line of thought"—is suggestive, and throws us back on a previous passage in the Address, in which we are told that "to construct the universe in idea it was necessary to have some notion of its constituent parts—of what Lucretius subsequently called the 'First Beginnings.'"[3]

The "First Beginnings!" What has "our present line of thought" to say on these? We shall do well to question it.

And, to begin at the beginning, we shall do well to note—not merely the order, but—the fact here admitted. There was—no matter when—an actual Beginning: a first start; distinct, definite. Antecedently, there was a prior time, when this first start had not been made. The process of Evolution, a "process of unravelling and unfolding," is a process which then had not begun. It is therefore not eternal. It had a beginning. But who began it?

[1] "Materialism and its Opponents," p. 598.
[2] "Belfast Address," p. 19.
[3] *Ibid.*, p. 2.

You postulate "Matter." But in so doing you are hypothecating a substance which before the "First Beginning" had not begun to be. How did it originate? Unable to answer that question, you make another assumption. You postulate "eternity" for that "matter" of whose origin you can give no account. But this accumulation of postulates will not help you. What *is* this matter which—impelled by the exigencies of Agnostic Evolution—you assume to have been self-originated? Make its essence what you will—extension, with Descartes; or palpableness, with Fechner—Matter is always, and is manifestly, the local lodgment, the objective manifestation, of Power. "The withered leaf is not dead and lost, there are Forces in it and around it, though working in inverse order; else how could it *rot?*"[1] Matter, Force, Motion, are not unknown to Science; but of matter self-originated and self-sustained, of matter self-existent and therefore eternal; of self-originated force, or self-originated motion; of all these throughout the realm of the inorganic world, Science knows nothing.

When therefore we have granted "the eternity of matter," the theory of Evolution is as far as

[1] Carlyle: "Sartor Resartus," book i. chap. xi. p. 43.

ever from being able to make a "beginning." That theory requires not merely matter, but matter in motion. Not merely matter in mass, but matter in its constituent atoms. Matter so minutely subdivided as to be immeasurably beyond the sphere of visibility; and yet matter not within the sphere of infinite divisibility. "The atoms" are "the first beginnings."[1] But speculation is at fault as to the mode in which, or the power by which, they "first began." In his panegyric on Lucretius, Professor Tyndall draws special attention to his "strong scientific imagination;"[2] and tells us that "his vaguely grand conception of the atoms falling eternally through space suggested the nebular hypothesis to Kant, its first propounder."[3] The "eternity" of these falling atoms, however, must not be confounded with the antecedent "eternity" of their origination. Like the "eternity" of the rhetorical preacher,[4] it has its own statute of limitations. It came to an end. While it lasted there might have been seen, "far beyond the limits of our visible world" (by aid of a

---

[1] "Belfast Address," p. 8.
[2] *Ibid*, p. 9.
[3] *Ibid.*, p. 10.
[4] Eternity: "An *infinite* candle; lighted—*at both ends*"!

"strong scientific imagination"), "atoms innumerable," "falling silently through immeasurable intervals of time and space."[1]

"Falling eternally through space:" "falling silently through immeasurable intervals:" but this eternal silence was broken by "great shocks of sound," "the mechanical shock of the atoms;"[1] and this eternal falling came to an end when "the interaction of the atoms"[2] came to a beginning. How came that beginning? Nothing more simple. At first, the atoms, silently falling, fell in parallel lines. After that they *began* to deflect from the perpendicular. Not all of them; nor all in the same direction: but only so many, and in so many directions as were necessary to produce "the mechanical shock," and "the interaction." But falling is motion, and matter is inert, and atoms in motion are atoms in which inertness has been overcome by a force external to themselves, and falling atoms are atoms gravitating towards a centre. What centre? and how originated? Why should atoms in motion have moved originally all in one direction? or why should they have ceased to do so? What, and whence, is that

[1] "Belfast Address," p. 10. [2] *Ibid.*, p. 8.

Force which first moved them,—which moved them in parallel lines,—which deflected them from the perpendicular,—as assumed by the hypothesis?

"It is certain," according to "the doctrine of Evolution," "that the existing world lay, potentially, in the cosmic vapour." But where it lay before the cosmic vapour existed, deponent saith not. "The fundamental proposition of Evolution" is, as we have seen, "that the whole world, living and not living, is the result of the mutual interaction, according to definite laws, of the forces possessed by the molecules of which the primitive nebulosity of the universe was composed."[1] Fundamental, however, as Professor Huxley declares it to be, it is very far indeed from "The First Beginning."

This "nebulosity was composed" of certain "molecules." But nebulosity is a state or condition; not a substance. Like the rigidity of an iron bar, or the malleability of gold-leaf, or the ductility of copper wire, "nebulosity" is a word not of matter, but of mode. It denotes a property, or it specifies a condition; but it does not distinguish, still less does it

[1] Professor Huxley: "Critiques and Addresses," p. 305.

define, a substance. It is characteristic of unintelligible hypotheses, not less than of "cosmic gas." In this instance, however, let it pass. We will not say that it was "caused," —that word might lead us back in the search for a *vera causa* to a "first beginning,"— but only that it was "composed." We will not even inquire who "composed" it. And yet, if it were permitted us to inquire at all, we might perhaps be excused for asking, How do you know that this nebulosity was "primitive"? or that its constituent "molecules" were "possessed" of forces? or that these forces were controlled by "definite laws"? or that the relation between them was that of "mutual interaction"? or "that the whole world, living and not living,"—the molecules themselves included,—"is the result" solely and exclusively of the "mutual interaction" which you have imagined?

What a tissue of conjectures is here And yet all this is assumed as "certain," and is postulated as "the fundamental proposition of Evolution."

But now, suppose it certain: what then? It leaves us as far as ever from a knowledge of "the first beginnings." It tells us of "forces" controlled by "definite laws." But if it tells us truly, then the law is the controlling Power, and has a priority over the powers controlled. Then "the forces possessed by the molecules" were at best subordinate and secondary: the "definite laws" alone were primary and supreme. But laws never make themselves. Who made these? and who made them thus distinctly "definite"?

But even their definiteness is not greater than their complexity. And this complexity —immeasurably beyond our power of exploration—is everywhere adjusted to the attainment of a common end. Who originated a complexity so intricate, yet so illimitable? Who established this unvarying adjustment of it—in the very "first beginning"? For we are now asked to imagine space filled with a frictionless fluid; to suppose that some portions of this fluid did somewhere, somehow, by some means, at some time or other, become "rotational;" and that having by rotation gained rigidity, they can now, by the latest triumphs of hydrodynamics, be "proved" to be indivisible and indestructible. Let it be granted. Granted that light, heat, sound, electricity, magnetism, are molecular movements mutually transmutable; that arrested molar movement displays itself as molecular movement; that the pressure of

Scientific Sophisms. 165

a gas is due to the varying motion of its molecules impinging on the walls of the vessel that contains it; that the rigidity, or space-occupying power of matter, is due to the formation of vortices in a frictionless ether, and that each vortex-atom is thenceforth indestructible; when the reality of the conceptions thus assumed has been granted, then by exactly so much has the absolute necessity been increased of assigning—at "the first beginning" —a First Cause, equal not only to the origination of Matter and of Force, but equal to the origination of matter thus constituted, and of force thus adjusted.

Evolution is thus seen to be the measure of Involution. Whatever has been evolved in the Effect was previously involved in the Cause. To deny this is to affirm that the effect may transcend the cause. If therefore—though in utter contempt of scientific verity—we were to resolve all chemical forces into forces mechanical, all life into chemistry, and the infinite diversity of living beings into mere variety in the play of molecular forces, ultimately resolving itself into a motion or motions of the universal ether, we should simply have increased by so much our previous estimate of the Power which—at the "first beginning"—was able thus "potentially" to endow "the cosmic vapour."

Matter, Force, Order, Law, Diversity in Unity, Concord in Complexity: they are all known to us, but not one of them is known as self-originated. Distinct in character, definite in operation, invariable in result: who made them so? You trace "the whole world, living and not living," to certain "properties" of Matter, acted upon by certain capacities of Force, operating in an invariable Order, under the reign of Law. You do well. Pursue your induction to "The First Beginnings." Whence came those "properties" of matter? those capacities of force? Order could not regulate them before Matter received them. Could Matter create them? Through all the "immeasurable intervals of time and space," Matter has never created one single atom. *Causa causarum:* what was that? Whatever it was, you will not be able to ignore it, except by refusing to go back to " The First Beginning."

That "first" beginning was followed by a second. Immovably based on the deep foundations of the inorganic world, there rises everywhere, elaborate and multifarious, the mysterious superstructure of organization and Life.

Scientific Sophisms. 167

No conclusion of modern science is more widely received or more confidently maintained than that which teaches that in the early history of our planet life was unknown. Not only was it not actual: it was not possible. Life then was not. But now life is. Life, then, had a beginning. What was that beginning? And whence?

"If," says Professor Huxley,[1] "the hypothesis of Evolution be true, living matter must have arisen from not-living matter, for, by the hypothesis, the condition of the globe was at one time such that living matter could not have existed on it, life being entirely incompatible with the gaseous state." And he adds that, even if we adopt Sir William Thomson's theory, that life on this planet may have been derived from life on some other, the difficulty of accounting for its origination is as great as ever. For the nebular theory, which is a part of the hypothesis of Evolution, asserts that all the worlds were once in "the gaseous state."

"But," he continues, "living matter once originated, there is no necessity for another origination, since the hypothesis postulates the unlimited, though perhaps not indefinite, modifiability of such matter." Waiving, for

[1] *Encyclopædia Britannica*, Article "Biology."

the present, the "unlimited modifiability" thus postulated, it is important to observe the profound significance of the admission here made. "Living matter once originated:" yes, but how? To that crucial question, the answer, on the same high authority, is given in these words: "Of the causes which have led to the origination of living matter, it may be said that we know absolutely nothing." "The present state of knowledge furnishes us with no link between the living and the not-living."[1] But however inscrutable the mode, there is no question—nor any room for question—as to the fact. "Living matter" was "once *originated*." Life had a Beginning.

Impenetrable, however, as is the veil which hides from our observation the origin of Life, still more inscrutable is the mystery which shrouds the first emergence of the self-conscious Mind.

Mr. John Stuart Mill admits the existence of the mind in the form of a "thread of consciousness," "aware of itself as past and future," and possessing a conviction of the simultaneous existence of other "threads of consciousness" and of numerous permanent possibilities of

[1] *Encyclopædia Britannica*, Article "Biology."

sensation.[1] And Professor Huxley asks, "Is our knowledge of anything we know or feel more or less than a knowledge of states of consciousness?" "And," he adds, "our whole life is made up of such states."[2] And again, in the same connection, he tells us of that "highest degree of certainty which is given by immediate consciousness."

But what then is this consciousness? and whence? Professor Huxley's language on the subject is particularly confident, although at present it is merely prophetic. "I hold," he says, "with the materialist, that the human body, like all living bodies, is a machine, *all* the operations of which will sooner or later be explained upon physical principles." And again: "I believe that we shall arrive at a mechanical equivalent of consciousness, just as we have arrived at a mechanical equivalent of heat."[3] But the vaticinatory character of these opinions is their least remarkable feature. Professor Huxley "holds" that all living things are machines, and "believes" that "thought is as much a function of matter as motion is;" but, as Dr. Beale observes, "of evidence in

[1] Mill upon Hamilton, p. 212.
[2] "Lay Sermons:" Descartes, p. 359.
[3] *Macmillan's Magazine*, vol. xxii. p. 78.

support of these beliefs there is none that will bear investigation, none that would convince any reasonable being." "Such opinions and beliefs on the mechanics of life and thought are certainly very striking, but it is remarkable that no one who entertains them has considered it necessary to adduce facts or arguments in their support. The mechanical theory of life and consciousness rests upon authority whose utterances are dogmatic and not dependent upon reason, fact, observation, and experiment."[1]

Widely different is the language of Mr. Herbert Spencer, and of Professor Tyndall, in which we are assured that "our states of consciousness are mere symbols of an outside entity which produces them and determines the order of their succession, but the real nature of which we can never know."[2] It must not be concealed however that, after all, Professor Tyndall does not differ from Professor Huxley more widely than Professor Huxley differs from himself. It is not always that he indulges in prophetic imaginings of "a mechanical equivalent of consciousness." When, as above quoted, he tells us of what he "holds" "with the materialist," we have only to turn to his "Phy-

[1] "Protoplasm; or Matter and Life." 1874. P. 119.
[2] "Belfast Address," p. 57.

siology" to find materials for the utter refutation of materialism.

"We class," he says, "sensations along with *emotions*, and *volitions*, and *thoughts*, under the common head of states of *consciousness*. But what consciousness is, we know not; and how it is that anything so remarkable as a state of consciousness comes about as the result of irritating nervous tissue is just as unaccountable as the appearance of the Djin when Aladdin rubbed his lamp in the story."[1]

"Some," says Dr. Beale, "have taught that mind transcends life, and life transcends chemistry, just as chemical affinity transcends mechanics. But no one has proved, and no one can prove, that mind and life are in any way related to chemistry and mechanics."[2] Even if the step from mechanics to chemistry had been admitted as ascertained and proved, it would still remain true that the step from chemistry to life is a mere unsupported assumption; an assumption "without the slightest reason."

"How any material impressions should awake thought; but, still more, how, in independence of all impressions, thought should be all the while there, alive and active, a world by itself—that

[1] P. 193.
[2] "Protoplasm," p. 299.

is the mystery." And that mystery no scalpel, no microscope, will ever explain. "Mechanical balances the most delicate, chemical tests the most sensitive, are all powerless there. And why? Simply because consciousness and they are incommensurable: of another nature, of another world from the first, sundered from each other by the whole diameter of being."

But whence came this "other world," this new "incommensurable"? and whence the "great gulf," the impassable chasm, which marks the new beginning? *Mens agitat molem;* but that implies for *Mens* a special nature, a special relation, and a special origin. What was that origin? and whence?

Whatever its source, whatever its nature, the one broad patent fact remains alike indubitable and incontestable:—there was a definite epoch in which the human mind first came into being. Thought began to be. Intelligence, self-conscious, emerged—though not from the world of matter—to be enthroned in the World of Mind. Whence came it? Who will tell us? For to Agnostic Evolution a phenomenon so portentous is absolutely fatal. Scientific Materialism can give no account of it. It is perfectly "UNACCOUNTABLE."

And yet it is true!

CHAPTER IX.

THE THREE BARRIERS.

"So long as you have that fire of the heart within you, and know the reality of it," says Mr. Ruskin, "you need be under no alarm as to the possibility of its chemical or mechanical analysis. The philosophers are very humorous in their ecstasy of hope about it; but the real interest of their discoveries in this direction is very small to human-kind."[1] And the same may be said of the discoveries themselves. Their actual amount, not less than their real interest, is "very small." So small indeed, that "their ecstasy about it"—though merely an "ecstasy of hope"—is a "very humorous" spectacle. He who doubts this has not read Mr. Darwin.

"It requires a long succession of ages to adapt an organism to some new and peculiar form of life, as, for instance, to fly through the

[1] "The Queen of the Air." London, 1869, p. 70.

air."[1] "We do not see the transitional grade through which the wings of birds have passed; but *what special difficulty is there in believing* that it might profit the modified descendants of the penguin, first to become enabled to flap along the surface of the sea, like the logger-headed duck, and ultimately to rise from its surface and glide through the air?"[2] "The tail of the giraffe looks like an artificially constructed fly-flapper; and *it seems at first incredible* that this should have been adapted for its present purpose by successive slight modifications, each better and better, for so trifling an object as driving away flies; yet we should pause before being too positive even in this case, for . . . a well-developed tail having been formed in an aquatic animal, it might subsequently come to be worked in for all sorts of purposes—as a fly-flapper, an organ of prehension, or as aid in turning, as with the dog."[3]

In this way, the tail of a horse may have been derived from that of a shark, the tail of a cow from the skate, and the giraffe owe his fly-flapper to a remote progenitor, the sturgeon. Or, if there be any who think that to affirm this

[1] "Origin of Species," First Edition, p. 328.
[2] *Ibid.*, p. 329.
[3] *Ibid.*, p. 215.

The Three Barriers. 175

is to affirm too much, Mr. Darwin may still ask (as above) "What special difficulty there is in *believing*" it? Especially "since it certainly is not true that new organs appear suddenly in any class."[1]

The counterpart of this strange story is still more worthy of a place in the record of the "Thousand and One Nights." For not only have so many terrestrial creatures been derived from an "aquatic origin"[2] by that marvellous metaphor called Natural Selection, but, on the other hand, there are not wanting some land-animals that, renouncing their original nature, have become aquatic. Surprising as it may be to learn that a giraffe was once a fish, it is not less surprising to be told that a whale was once a bear, And yet, " In North America, the black bear was seen by Hearne swimming for hours with widely-open mouth, thus catching, like a whale, insects in the water. *I see no difficulty* in a race of bears being rendered by Natural Selection more and more aquatic in their structure and habits, with larger and larger mouths, till a creature was produced as monstrous as a whale."[3] With this difference, however: that,

[1] "Origin of Species," First Edition, p. 214.
[2] *Ibid.*, p. 215.
[3] In the third and subsequent editions, the latter part

when the ursine whale began his career he had his tail to make—an operation exactly the reverse of that in the previous story. The land animals, having been fishes, derived their tails from the waters; but in this latter case a land animal goes into the water to live like a fish and procure a tail. Humorous? Not at all. Perfectly serious. Consider the authority of Mr. Huxley, and remember that "the hypothesis postulates the unlimited modifiability of matter."

Nor is it matter alone which, in the hands of "Natural Selection" presents the marvellous transformations due to unlimited modifiability. "Under changed conditions of life," says Mr. Darwin, "it is at least possible that slight modifications of instinct might be profitable to a species; and if it can be shown that instincts do vary ever so little, then *I can see no difficulty* in Natural Selection preserving and continually accumulating variations of instinct to any extent that was profitable. It is thus, *I believe*, that all the most complex and wonderful instincts have originated."[1]

of this passage is omitted, for no apparent reason. No hint is given that Mr. Darwin now sees any difficulty where he saw none before, and the statement as now left still contains the suggested transformation; a suggestion strengthened by the connection in which it is found.

[1] "Origin of Species," p. 229.

The Three Barriers.

This is too much for M. Flourens. "Surely," says that accomplished naturalist, "we cannot take this as meant to be serious. Natural Selection choosing an instinct!

> '. . . La poésie a ses licences, mais
> Celle-ci passe un peu les bornes que j'y mets.'"[1]

Mr. Darwin, however, is serious enough, and maintains in all good faith, that peculiar instincts are in all cases the result not of original endowment, but of subsequent acquisition ; "by the slow and gradual accumulation of numerous slight, yet profitable variations."[2] Individual life, as well as the life of the community, whether in ants or bees, was once a totally different thing from what we now behold ; then beavers did not build, and neither the stork nor the swallow knew their appointed seasons.

In treating of the ants and the honey-bee, Mr. Darwin attempts to account for that striking peculiarity—the groundwork of much of their polity—the existence of neuters.

"Thus, *I believe*," he says, "it has been with social insects ; a slight modification of structure *or instinct*,

---

[1] "Examen du Livre de M. Darwin sur L'Origine des Espèces." Par P. Flourens. (Paris, 1864.) P. 55. *Vide infrà:* Appendix, Note C.
[2] "Origin of Species," p. 230.

correlated with the sterile condition of certain members of the community, has been advantageous to the community; consequently the fertile males and females of the same community flourished, and transmitted to their fertile offspring a tendency to produce sterile members, having the same modification. And *I believe* this process has been repeated, until that prodigious amount of difference between the fertile and sterile females of the same species has been produced, which we see in many social insects."[1]

But the very existence of "the community" (in the case of the honey-bees, for example) depends upon the specific arrangements of the present polity and constitution. Alter these arrangements, and the polity is at an end; "the community" exists no longer. If, therefore, at any time, all the females were fertile, as this explanation implies that they once were, then "the community" did not exist; and its operations, however "slight," in "modification of structure, or instinct," at a time when it was non-existent, are unimaginable, except in Utopia.

If only they were imaginable, the "scientific imagination" would not lack exercise. We should in that case have to imagine that when the fertile females were transforming—not themselves but—their posterity into sterile members

[1] "Origin of Species," p. 260.

for the benefit of society, there was one remarkable exception. One female there was who, by a long preconcerted scheme, though by a most occult and undiscoverable process, was all the while prodigiously increasing her fertility in order to become the sole Mother and Queen of the whole hive! We should have to imagine fertile animals agreeing to produce, and actually producing, sterile offspring! "The fertile males and females flourished; and transmitted to their fertile offspring a tendency to produce sterile members!" Fertile parents transmit, through fertile progeny, a tendency to produce sterility incapable of further production! "Humorous"? Not at all. The theory requires it, and therefore, quite seriously, Mr. Darwin "believes it."

By one of his earliest and acutest critics it was justly observed, that "If we except a passing cavil at the imperfect knowledge of optics displayed in the mechanism of the eye, Mr. Darwin can scarcely be said to have touched the evidence for design deduced from the felicities of fabric and deep-lying adjustments, so profusely exemplified throughout the animal kingdom. He tells us indeed how the pigeon's feather may be varied, but not how the pigeon came to be feather-clad at all. He leaves us quite in the dark also as to the mode in which natural selection sets to work in the multiplying of air-sacs, or in the boring of bones, to increase the facilities for flotation and flight. But he devotes a large portion of a chapter on Instinct, otherwise extremely graceful and interesting, to a hypothetical exposition of the processes by which the common hive-bee, *Apis mellifica*, *might* have distanced her less skilful kindred *Melipona* and *Bombus;* and how the wonderful phenomena of sexual suppression and vicarious labour *might* have arisen among the social instincts of the bee and ant tribes generally. No one, since Touchstone's time, has set such store on the virtues, or so taxed the capacities, of an *If*. A certain abstract theorem conceded, *if* Bombus or Melipona could be brought to put that theorem in practice, one huge stumbling-block would be removed from Mr. Darwin's speculative path. But this is the hitch. It is as much out of the question for Bombus or Melipona, not being a man, to see with Mr. Darwin's eyes, as it would be for Mr. Darwin, not being a bee, to work with Melipona's tools. Slight deflexions of habit, artificially provoked, in the more highly endowed insect, do not furnish the smallest presumption of the genesis of new endowments in its inferior sisterhood.

'It may easily be *supposed*,' in these researches, is but a sorry substitute for, 'It has actually been *observed*.' The true tokens of consummate geometrical prescience can never be simulated by tentative effort. Had Mr. Darwin lived two thousand years ago, his ceral experiments might have furnished a target for the shafts of Aristophanes;[1] but, indifferent alike to savant and satirist, Melipona was then building her cells no better, and Mellifica no worse. Those explanations of the mystery of cell-making which really explain nothing are, however, moderation itself to the inimitable though unconscious legerdemain which converts an unanswerable and unblunted objection to our author's favourite solvent, drawn from the phenomena of insect sterility and caste, into the occasion of a panegyric on its power. It is his business to prove that natural selection *has done* certain wonderful things: See, he virtually says, what wonderful things, far beyond my own expectation, natural selection *can do*.[2] A more flagrant intrusion of unpruned fancy into a domain sacred to the severities of observation can scarcely be conceived.

"The social insects, like those lower in the

[1] "Clouds," 147-153.
[2] "Origin of Species," p. 242.

scale, must have started, on Mr. Darwin's view, as ordinary male and female, with a common share of individual labour; on a par, in this respect, with a flock of geese, or a herd of cattle, or a community of mankind. Now let any breeder of cattle consider through what agencies a variety could be attained of which only one birth in five should be a bull or a cow, the other four being natural neuters, devoted subjects of their perfect sister, but sworn foes of her spouse. It is an aptitude precisely analogous to this that has produced, we are asked to believe, the economy of the bee-hive. Or let any transatlantic admirer of the 'domestic institution' of *Formica rufescens*, turn over in his mind the means by which every third man-child born on his estate should be ten times the size of the rest of the family;[1] or each alternate female be fitted for a nurse while forbidden to be a

---

[1] Mr. Darwin, in noting the fact that "the neuters of several ants differ, not only from the fertile females and males, but from each other, sometimes to an almost incredible degree," says, "The difference between them is the same as if we were to see a set of workmen building a house, of whom many were five feet high, and many sixteen feet high—but we must further suppose that the larger workmen had heads four times as big as those of the smaller men, and jaws nearly five times as big."—"Origin of Species," pp. 260, 261

mother; and he would have the measure of the intrinsic likelihood of the Darwinian doctrine, in its bearing on that insect and its confederates. It were idle to enlarge. There are worthier lessons to be gleaned from the world of instinct than such as affront all legitimate analogy, and gratuitously dissociate the marvels of nature from their only true solvent, the ordination of God."

Turning now from the disordered dreams of unpruned fancy to the severities of observation; from ingenious suppositions of what might have been, to the actual certainties that are; we find all Comparative Anatomy tending towards the recognition and extrication of three supreme values, in the grouping of animals, and the graduation of life, past as well as present :— the BACKBONE, the BREAST, and the BRAIN. And the key to the significance of animal life and its prerogatives, thus grouped and graduated, is not, and cannot be, Selective Development, but is, and must be, Elective Design.

"The first leet, in the ascending order, takes note of all animals, as Vertebrates or Sub-vertebrates: for every individual organism endowed with a backbone, there are countless millions without it. Hence this first or exterior leet leaves a master-group, palpably supreme in framework and ground-plan over three other groups—the Articulate, the Convolute, and the Radiate—between which and the master-group the BARRIER OF BACKBONE stands impassable; at least till it is explained how a butterfly could become a bird, or a snail a serpent, or a starfish acquire the skeleton of the salmon or the shark. It is like the going forth of a Divine decree: 'One shall be taken, and three shall be left.'

"The second leet, Sub-vertebrates out of view, takes account of Vertebrates themselves as Mammals or Sub-mammals. Among the elect it makes an inner election. Besides the Backbone it exacts the Breast; shedding off, as before, three well-marked groups subordinate to the master-group of Mammals or Sucklers. These breastless tribes are Birds, Reptiles, and Fishes; holding high, low, and medium rank among themselves, not so much on the principle of skeleton, or its specialized offshoots, as on that of characters which are correlated to the development of care for their young. . . . Still the Mammal, by its endowment of the fostering bosom, stands elect, aloft, and apart— Bird, Reptile, Fish, far beneath in the scale;

The Three Barriers.

and till it is shown how an animal that never *got* suck stumbled on the capacity of *giving* what was never given it, the BREAST will stand, against all dreams of development, COMPANION-BARRIER to the Backbone. Again is heard the elective edict: 'One shall be taken, and three shall be left.'

"Third, last, innermost leet: note has to be taken among the Mammalia themselves, from the Marsupials to Man, of the presence or absence of one testing character, and that the chief—the Perfect Brain. This is found in one creature, occupying, as it were, the inner ring and core of the concentric circles of vitality, and in one alone. In the lowest variety of man it is present—present in the Negro or the Bushman as in the civilized European; and absent in *all below* man—absent in the ape or the elephant as truly as in the kangaroo or the duckmole. To *all* men the pleno-cerebral type is *common:* to *man*, as such, it is *peculiar*. And till we hear of some Simian tribe which speculates on its own origin, or discusses its own place in the scale of being, we shall be safe in opposing the HUMAN BRAIN, with its sign in language, culture, capacity of progress, as BARRIER THE THIRD to Mr. Darwin's scheme."[1] "And thus, as in the former leets, are the triple tribe of under-brains walled off from the Brain of Man.[2] A third time there falls a voice from the Excellent Glory: 'One shall be taken, and three shall be left.'"

Below the fish, how *powerless* comparatively, all creatures are! The primates of sub-vertebrate nature are the ant and the bee. Most mollusks are anchored to one spot for life, and the bulkiest of crustaceans, shorn of other locomotion, could only crawl in shallow waters among his rocks and sands. The advent of the backbone is the advent of animal power: the type of an all-pervading and resistless energy. The wing of the eagle, the jaw of

---

[1] "The Three Barriers: Notes on Mr. Darwin's 'Origin of Species.'" Blackwood & Sons. Pp. 88, *et seqq*. To the highly-gifted author of this brilliant little book—a book as admirable in method as unanswerable in effect—I gratefully record my obligations.

[2] "By a purely inductive process, the sub-human mammalia have been carefully distributed into the wave-brained, the smooth-brained, and the loose-brained, represented respectively by the ape, the beaver, and the kangaroo; with a result, so far as the two departments of science are comparable, like that of the application of Kepler's laws to the planetary motions: the subjects of the classification fall, for the first time, into their true places—a mob of animals becomes a regular army."

the crocodile, the spring of the tiger, the teeth of the shark, the terrible coil of the boa-constrictor; the backbone is the basis of them all.

Below the mammal, again, how *loveless*, by comparison, is the world of life! There are no sub-mammalian mothers; animals below that line are parents or producers only. The crossing of that line is a great work of Deity. God creates a new thing in the earth when He hangs the nursling on the mother's breast, and bids the two be as one. Together with the prerogative of the nurturing bosom there start up everywhere, on land and sea, the most touching examples of brute devotion and of passionate maternity.

Deep calleth unto deep, and the cry is still *Excelsior!* Nature is a hierarchy of which the head is man. Mind, language, worship, civilization; the will to determine, the tongue to speak, the hand to do; these—in their boundless purport—are all lacking until the Creator plants upon the scene the solitary owner of the Perfect Brain. Named in one word, all these are *wisdom;* and Man, "thinker of God's thoughts after Him," is, among uncounted myriads of lower existences, on this earth, Only Wise. Of this superiority, the human brain is the badge. The attempts that have been made to minimize, and even to efface its significance, will be noticed in the sequel; but the force and effect of that significance are not to be invalidated and cannot be impaired by disputations in detail. The one broad characteristic fact remains beyond dispute: all healthy human brains are structurally perfect; but the highest brute brains are structurally imperfect. The human brain is pleno-cerebral; all other brains are manco-cerebral. The human brain, in its least cultivated manifestations, retains the latent franchise of progressive reason; all other brains exhibit the rigid circumscription of unprogressive instinct. No brute is susceptible of human culture; while, on the other hand, of that culture there is no human infant that is not susceptible. Between these two, the difference thus seen is nothing less than a difference absolutely immeasurable.

CHAPTER X.

ATOMS.

BUT these magnificent achievements — the Vertebral Column, the Fostering Bosom, the Perfect Brain—with their inexplicable origin, their profound significance, their limitless results, have been accomplished by the cosmical atoms alone. Outside those atoms, or beyond them, there is not now, nor has there been at any time, any existence whatever. No substance, no essence, no entity, no force, no motion. "Matter is the origin of all that exists; all natural and mental forces are inherent in it."[1] "The existing world lay potentially in the cosmic vapour."[2] For "the fundamental proposition of evolution" is, as we have seen, "that the whole world, living and not living, is the result of the mutual interaction, according to definite laws, of the forces possessed by the molecules of which the primitive nebulosity of the universe was com-

[1] Buchner, *ut sup.*, p. 96.
[2] Prof. Huxley, *ut sup.*, p. 64.

posed."[1] In a word—and that, the word of Lucretius, adopted and adorned in the Belfast Address—" The Atoms are the first beginnings."

What then are these ultimate inorganic atoms on which (according to the hypothesis of Development) everything depends? The idea expressed by the word itself is simply the idea of "matter" *in minimis*, arising only from an arrest by a supposed physical limit, of a geometrical divisibility possible without end. But "things which cannot be cut" might be all alike; or they might be variously different, *inter se;* and, on setting out in this inquiry it is necessary to know on which of these two assumptions we are to proceed. If the materialist is to be credited with any logical exactness, it is the former assumption alone that is admissible. When he asks for *no more than matter* for his purpose of constructing a universe, his demand is restricted to *the essentials fo matter*, the characters which enter into its definition. It is from these alone that he pledges himself to deduce all the accessory characters which appear in one place though not in another, and which discriminate the several provinces of nature. It is in perfect

---

[1] Prof. Huxley, *ut suprà*, p. 64. *Vide infrà*, Appendix, Note J.

accordance with this, that the "atomists," says Lange, "attributed to matter only the simplest of the various properties of things — those, namely, which are indispensable for the presentation of a something in space and time, and their aim was to evolve from these alone the whole assemblage of phenomena." "They it was," he adds, "who gave the first perfectly clear notion of what we are to understand by matter as the basis of all phenomena. With the positing of this notion, materialism stood complete, as the first perfectly clear and consequent theory of all phenomena."[1]

If further corroboration of this statement were needed, it might be adduced from Mr. Herbert Spencer's definition of Evolution, already quoted:[2]—"Evolution is a change from an indefinite incoherent homogeneity, to a definite coherent heterogeneity, through continuous differentiations and integrations." And again: —"From the earliest traceable cosmical changes down to the latest results of civilization we shall find that the transformation of the homogeneous into the heterogeneous is that in which evolution essentially consists." In perfect consistency with these statements Mr. Spencer further

[1] "Geschichte des Materialismus," i. pp. 8, 9.
[2] *Vide suprà*, p. 27.

contends that the properties of the different elements (*i.e.*, the chemical elements, hydrogen, carbon, etc.) "result from differences of arrangement, arising from the compounding and recompounding of *ultimate homogeneous units*."[1] So that, to sum up all in one word, there is but, as he further tells us, "one ultimate form of matter, out of which the successively more complex forms of matter are built up."[2]

These statements are not lacking, either in clearness or consistency. Their only fault is that they are not correct. The "one ultimate form of matter" is not forthcoming. The "homogeneous extended solids" are not homogeneous. We are not to be surprised if we should see sixty-two out of the sixty-three "elements" fall to pieces analytically before our eyes. If we would speak positively of the simplicity of phosphorus or carbon, we are warned that "there are no recognised elementary substances, if the expression means substances known to be elementary. What chemists for convenience call elementary substances, are merely substances which they have thus far failed to decompose."

But let the contrary supposition be admitted.

[1] *Contemp. Rev.*, June, 1872.
[2] "Principles of Psychology," vol. i. p. 155.

Atoms. 193

Let it be supposed that the alleged homogeneity were as real as now it is imaginary. Let the appeal be allowed which all logical atomists make to the case of *isomeric* bodies, and especially to that of *allotropic* varieties. Let such varieties of appearance as those presented by carbon[1] and phosphorus[2] be attributed, not to any qualitative cause, but to a different grouping of the atoms; the morphological differences, if adequately obtained, will still contribute no explanation of the observed variations of attribute. Vary in imagination, as you please, the adjustments of their homologous sides, so as to build molecules of several types, the question will still remain unanswered,—" What is there in the arrangement *a b c* to occasion 'activity' in phosphorus, while the arrangement *b a c* produces 'inertness?'" Where the products differ only in geometrical properties, and consequently in optical, the explanation may be admissible, the form and the laying of the bricks determining the outline and the density of the structure. But by no device can the deduction be extended from the physical to the

---

[1] Charcoal, black-lead, and diamond.
[2] In the yellow, semi-transparent, inflammable form; and again as an opaque, dark red substance, combustible only at a much higher temperature.

O

chemical properties : to these last heterogeneity is essential. To deduce chemical phenomena from mechanical conditions, if it be not an impossible conception, may possibly be a " figment of the intellect," but it is a figment without any pretence to " verification."

"Even in the last resort, if we succeed in getting all our atoms alike, we do not rid ourselves of an unexplained heterogeneity ; it is simply transferred from their nature as units to their rules of combination. Whether the qualitative difference between hydrogen and each of the other elements is conditional upon a distinction of kind in the atoms, or on definite varieties in their mode of numerical or geometrical union, these conditions are not provided for by the mere existence of homogeneous atoms ; and nothing that you can do with these atoms, within the limits of their definition, will get the required heterogeneity out of them. Make them up into molecules by what grouping or architecture you will ; still the difference between hydrogen and iron is not that between one and three, or any other number ; or between shaped solids built off in one direction and similar ones built off in another, which may turn out like a right and a left glove. If hydrogen were the sole 'primordial,' and were transmutable, by select shuffling of its atoms, into every one of its present sixty-two associates, both the tendency to these special combinations, and the effects of them would be as little deducible from the homogeneous datum as, on the received view, are the chemical phenomena from mechanical conditions. I still think, therefore, that if you assume atoms at all, you may as well take the whole sixty-three sorts in a lot. And this startling multiplication of the original monistic assumption, I understand Professor Tyndall to admit as indispensable."[1]

This witness is true. The "original monistic assumption" is now discarded by Professor Tyndall[2] and Professor Bain as emphatically as by Mr. Martineau himself. The "ultimate homogeneous units" of Mr. Spencer are now found to be utterly inadequate to the task required of them. They must be in motion; they must be of various shapes; they must be of as many kinds as there are chemical elements; for how could we possibly get water if there were only hydrogen elements to work with? And when, by means of this very considerable enlargement of the original datum we have got water, what is that further enlargement by which we should be able so to manipulate our ever-increasing resources as to educe, for example, consciousness? Let some Power so ordain, and some Wisdom so contrive, that all the atoms are affected by gravitation and polarity; let there be, as Fechner insists that there is, a difference among molecules; let there be the *inorganic*, which can change only their *place*, like the

---

[1] The Rev. James Martineau: *Modern Materialism* (*Contemp. Rev.*, vol. xxvii. p. 338).
[2] For Prof. Bain's *dicta*, see "Mind and Body": pp. 124-135.

particles in an undulation; and the *organic*, which can change their *order*, as in a globule that turns itself inside out. What then? When you have to pass from mere sentiency to thought and will, your Theory of Development is as impotent as ever until you have obtained—what only a further hypothesis can give—a handful of Leibnitz's monads to serve as souls in little.

"But surely you must observe that this 'matter' of yours alters its style with every change of service; starting as a beggar, with scarce a rag of 'property' to cover its bones, it turns up as a prince when large undertakings are wanted . . . It is easy travelling through the stages of such a hypothesis; you deposit at your bank a round sum ere you start, and drawing on it piecemeal at every pause, complete your grand tour without a debt."[1]

If now, from fictitious fancies such as these, we turn to the actual facts, we shall find that the whole argument sums itself up in a single remark of Sir William Thomson: "The assumption of atoms can explain no property of body which has not previously been attributed to the atoms themselves."

[1] Martineau: "Religion as Affected by Modern Materialism."

The "atom" of the modern mathematical physics has, accordingly, given up its pretension to stand as an absolute beginning, and now serves only as a necessary rest for exhausted analysis, before setting forth on the return journey of deduction. "A simple elementary atom," says Professor Balfour Stewart, "is probably in a state of ceaseless activity and change of form, but it is, nevertheless, always the same."[1] "The molecule," (here identical with "atom," as the author is speaking of a simple substance, as hydrogen), "though indestructible, is not a hard rigid body," says Professor Clerk Maxwell, "but is capable of internal movements, and when these are excited it emits rays, the wave-length of which is a measure of the time of vibration of the molecule."[2] But "change of form" and "internal movements" are impossible without shifting parts, and altered relations; and where then is the final simplicity of the atom? It is no longer a pure unit, but a numerical whole. And as part can separate from part, not only in thought but in the phenomena, how is it an "atom" at all? "What is there, beyond an arbitrary dictum, to prevent a part which

[1] "The Conservation of Energy," p. 7.
[2] "A Discourse on Molecules," p. 12.

changes its relation to its fellows from changing its relation to the whole—removing to the outside? Such a body, though serving as an element in chemistry, is mechanically compound, and has a constitution of its own, which raises as many questions as it answers, and wholly unfits it for offering to the human mind a point of ultimate rest. It has accordingly been strictly kept to a penultimate position in the conception of philosophical physicists like Gassendi, Herschel, and Clerk Maxwell, and of masters in the logic of science, like Lotze and Stanley Jevons."

Nor is it to be overlooked that the sixty-three kinds of atoms are not at liberty to be neutral to one another, or to run an indeterminate round of experiments in association, within the limits of possible permutation. "Each is already provided with its select list of admissible companions; and the terms of its partnership with every one of these are strictly prescribed; so that not one can modify, by the most trivial fraction, the capital it has to bring. Vainly, for instance, does the hydrogen atom, with its low figure and light weight make overtures to the more considerable oxygen element; the only reply will be, either none of you or two of you. And so on throughout the list." It is

In view of this property of admitting certain definite possibilities, while yet they are so limited as to fence off and exclude all competing possibilities, that Sir John Herschel felt himself compelled to describe the atoms as possessing *" all the characteristics of manufactured articles."*

This verdict amuses Dr. Tyndall; nothing more. " He twice[1] dismisses it with a supercilious laugh; for which perhaps, as for the atoms it concerns, there may be some suppressed '*ratio sufficiens.*' But the problem thus pleasantly touched is not one of those which *solventur risu;* and, till some better grounded answer can be given to it, that on which the large and balanced thought of Herschel and the masterly penetration of Clerk Maxwell have alike settled with content, may claim at least a provisional respect."[2]

To conclude. The conception of an infinitude of discrete atoms, when pushed to its hypothetical extreme, brings them no nearer to unity than *homogeneity,*—an attribute which itself implies that they are separate and comparable members of a *genus.* And what is the result of comparing them?

[1] Belfast Address, p. 26; and *Fortnightly Review,* Nov., 1875, p. 598.
[2] Martineau: *Contemporary Review,* vol. xxvii. p. 345.

They "are conformed," we are assured, "to a constant type with a precision which is not to be found in the sensible properties of the bodies which they constitute. In the first place, the mass of each individual," "and all its other properties, are absolutely unalterable. In the second place, the properties of all" "of the same kind are absolutely identical."[1] Here then, to adopt the weighty words of Mr. Martineau, "we have an infinite assemblage of phenomena of Resemblance. But further, these atoms, besides the internal vibration of each, are agitated by movements carrying them in all directions, now along free paths, and now into collisions.[2] Here therefore, we have phenomena of Difference in endless variety. And so it comes to this: that our unitary datum breaks up into a genus of innumerable contents, and its individuals are affected both with ideally perfect correspondences and with numerous contrasts of movement. What intellect can pause and compose itself to rest in this vast and restless crowd of assumptions? Who can restrain the ulterior question,—WHENCE then these myriad types of the same letter imprinted on

[1] Prof. Maxwell's "Discourse on Molecules," p. 11.
[2] "Theory of Heat." By J. Clerk Maxwell, M.A., LL.D., F.R.SS. London and Edinburgh. Pp. 310, 311.

the earth, the sun, the stars, as if the very mould used here had been lent to Sirius, and passed on through the constellations?"

For answer to this "ulterior question," we shall find none more conclusive, none more authoritative, than that of Professor Maxwell :—

"No theory of evolution can be formed to account for the similarity of the molecules throughout all time, and throughout the whole region of the stellar universe, for evolution necessarily implies continuous change, and the molecule is incapable of growth or decay, of generation or destruction."

Again he says: "None of the processes of Nature, since the time when Nature began, have produced the slightest difference in the properties of any molecule. On the other hand, the exact equality of each molecule to all others of the same kind precludes the idea of its being eternal and self-existent. We have reached the utmost limit of our thinking faculties when we have admitted that because matter cannot be eternal and self-existent it must have been created."

"These molecules," he adds, "continue this day as they were created, perfect in number, and measure, and weight; and from the ineffaceable characters impressed on them we may learn that those aspirations after truth in statement, and justice in action, which we reckon among our noblest attributes as men, are ours because they are the essential constituents of the image of Him, who in the beginning created not only the heaven and the earth, but the materials of which heaven and earth consist."[1]

A fit pendant to this noble utterance is furnished in the words of Professor Pritchard, who, quoting this passage, adds,—

"And this is the true outcome of *the deepest, the most exact, and the most recent science of our age.* A grander utterance has not come from the mind of a philosopher since the days when Newton concluded his Principia by his immortal scholium on the majestic Personality of the Creator and Lord of the Universe."[2]

[1] "Discourse on Molecules."
[2] *Address* at the Brighton Congress, October, 1874.

CHAPTER XI.

APES.

"IF," says Prof. Tyndall, addressing his Birmingham audience, "If to any one of us were given the privilege of looking back through the æons across which life has crept towards its present outcome, his vision would ultimately reach a point when the progenitors of this assembly could not be called human. From that humble society, through the interaction of its members and the storing up of their best qualities, a better one emerged; from this again a better still; until at length, by the integration of infinitesimals through ages of amelioration, we came to be what we are to-day."[1]

If we ask for some warrant of evidence in support of this series of assertions founded on assumption, Mr. Darwin replies that "On the principle of Natural Selection with di-

[1] "Science and Man:" *Fortnightly Review*, 1877, p. 116.

vergence of character, it does not seem incredible, that from some such low and intermediate form as the lower algæ, both animals and plants may have been developed; and if we admit this, we must admit that all organic beings which have ever lived on this earth may have descended from some one primordial form."[1]

In other words, and to speak more precisely, "Born of Electricity and Albumen, the simple monad is the first living atom; the microscopic animalcules, the snail, the worm, the reptile, the fish, the bird, and the quadruped, all spring from its invisible loins. The human similitude at last appears in the character of the monkey; the monkey rises into the baboon; the baboon is exalted to the ourang outang; and the chimpanzee, with a more human toe and shorter arms, gives birth to Man."[2]

What Sir David Brewster has here done for the *Fauna* on this principle of Development, Hugh Miller has in like manner done for the *Flora*, when he tells us that according to this theory "dulse and hen-ware became, through a very wonderful metamorphosis,

---

[1] "Origin of Species:" p. 519.
[2] "North British Review," 1845, p. 483.

cabbage and spinach; that kelp-weed and tangle burgeoned into oaks and willows; and that slack, rope-weed, and green-raw, shot up into mangel-wurzel, rye-grass and clover."[1]

And all this—in Mr. Darwin's opinion—"does not seem incredible." There must have been—we have his word for it—"a series of forms graduating insensibly from some ape-like creature to man as he now exists."[2]

How to derive the "ape-like creature" himself? By a similar process:—"a series of forms graduating insensibly" from a tadpole to a monkey. The Ape is the immediate, but the Ascidian is the remote progenitor of the genus *Homo*. And these Ascidians, which "resemble tadpoles in shape, and swim by means of a vibratile tail, which they shake off when they quit the larva state and assume the sessile condition," "have been recently placed, by some naturalists, among the Vermes or worms."

As to the ape-like creature,—

"Man is descended from a hairy quadruped, furnished with a tail and pointed ears, probably arboreal in its habits, and an inhabitant

[1] "Footprints of the Creator," p. 226.
[2] "Descent of Man:" vol. i. p. 235.

of the old world."[1] And again :—"The early progenitors of man were no doubt well covered with hair, both sexes having beards; their ears were pointed and capable of movement; and their bodies were provided with a tail, having the proper muscles. . . . The males were provided with great canine teeth, which served them as formidable weapons."[2]

Then as to the Ape's descent from his Ascidian ancestor:—

"The most ancient progenitors in the Kingdom of the Vertebrata at which we are able to obtain an obscure glance, apparently consisted of a group of marine animals, resembling the larvæ of existing Ascidians. These animals probably gave rise to a group of fishes, as lowly organized as the Lancelet; and from these the Ganoids and other fishes like the Lepidosiren, must have been developed. From such fish a very small advance would carry us on to the amphibians. . . . Birds and reptiles were once intimately connected together, and the Monotremata now, in a slight degree, connect mammals with reptiles. But no one can at present say by what line of descent the three higher and

[1] "Descent of Man," vol. ii. p. 389.
[2] *Ibid.*, vol. i. pp. 206, 207.

related classes, namely, mammals, birds, and reptiles, were derived from either of the two lower vertebrate classes, namely amphibians and fishes. In the class of mammals the steps are not difficult to conceive which led from the ancient Monotremata to the ancient Marsupials; and from these to the early progenitors of the placental animals. We may thus ascend to the Lemuridæ; and the interval is not wide from these to the Simiadæ. The Simiadæ then branched off into two great stems, the New World and the Old World monkeys; and from the latter, at a remote period, man, the wonder and glory of the universe, proceeded. . . . . If a single link in this chain had never existed, man would not have been what he now is. Unless we wilfully close our eyes, we may, with our present knowledge, approximately recognize our parentage, nor need we feel ashamed of it."[1]

"If a single link in this chain had never existed"! Why, even as Mr. Darwin has imagined it, it is not a chain at all. There is no continuity of concatenation. Even its very first link has to be imagined. And even when it has been imagined it is found to consist—not really, not demonstrably, but

[1] " Descent of Man," vol. i. pp. 212, 213.

only—"apparently" "of a group of marine animals." Of this group we have no other view than a mere "glance,"—"an obscure glance." But this first link, even when on the strength of an obscure glance it has been pronounced "apparent," is still not even "apparently" connected with any other. The connection required by the hypothesis—very far indeed from being "apparent"—is "probable" only. "These animals *probably* gave rise to a group of fishes," "and from these the Ganoids and other fishes *must have been* developed." But why "must have been?" there is no sort of necessity except that which is due to the exigencies of the theory. "From such fish a very small advance would carry us on to the amphibians." Possibly: but this very small advance is not to be had. Mr. Darwin's argument is made by himself to depend on the strength of his "chain"; and the strength of his chain is precisely that of its weakest link. But before all questions of strength there must be the prior fact of existence. Chains are made not by an aggregation of detached links, but by their continuity of concatenation. "A very small advance,"— possibly: but to advance at all without the aid of the missing link, is to abandon the pretence of a chain. Yet this is precisely Mr. Darwin's chosen mode of progression.

"In the class of mammals," he tells us, "the steps are not difficult to conceive which led from the ancient Monotremata to the ancient Marsupials; and from these to the early progenitors of the placental animals." In this theory of Ascensive Development "the steps" are every thing. But where are they? Their discovery is hopeless, their demonstration is impossible; no matter: they "are not difficult to conceive"!

"We may thus ascend to the Lemuridæ." "Thus": by steps which cannot be found; steps on which no one ever stood; but still, steps which Mr. Darwin finds it "not difficult to conceive." And then: "from these to the Simiadæ" "the interval is not wide." So be it: but however it be, it is nothing to the purpose. That which is to the purpose is not the width, but the fact of "the interval." And this fact of "the interval" is attested by Mr. Darwin himself. And with this "interval" before him, and these aerial "steps," and these appearances which are "apparent" only to "an obscure glance," Mr. Darwin can so far overlook the obvious and actual, in his zeal for the ideal and imaginary, as to say—"If a single link in this chain had never existed!"

„Even this is not the worst. For, he adds "Unless we wilfully close our eyes, we may, with our present knowledge, approximately recognize our parentage." "Our present knowledge"! Why, that is merely our present want of knowledge; for it is he himself who tells us that "no one can at present say by what line of descent the mammals," *i.e.*, ourselves "were derived."

In the hands of Prof. Huxley, the specious plausibilities of Mr. Darwin commonly assume the form of dogmatic affirmations; but in relation to this matter of the Descent of Man from the Ape, the cautious and conditional generalisations of Mr. Huxley furnish fresh proof, if fresh proof be needed, of the thoroughly conjectural character of Mr. Darwin's theory.

"If," says the learned Professor, "IF Man be separated by no greater structural barrier from the brutes than they are from one another—THEN it SEEMS to follow that IF any process of physical causation can be discovered by which the genera and families of ordinary animals have been produced, that process of causation is amply sufficient to account for the origin of Man. In other words, IF it could be shown that the Marmosets, for example, have arisen by gradual modification of the ordinary Platyrhini, or that both Marmosets and Platyrhini are modified ramifications of a primitive stock—THEN, there would be no rational ground for doubting that man MIGHT have originated, in the one case, by the gradual modification of a man-like ape ; or, in the other case, as a ramification of the same primitive stock as those apes."[1]

Widely different from Mr. Darwin's "chain," with every "single link" in its place, this reiterated relation of "If" and "then"; with its conditional sequence of what, after all, only "seems to follow"; and its ultimate conclusion that "Man *might* have originated," either in a given mode, or in some other mode not given.

Mr. Huxley adds, " I adopt Mr. Darwin's hypothesis therefore, subject to the production of proof that physiological species MAY be produced by selective breeding."[2] But this desiderated "proof" is precisely that very thing concerning which both Mr. Huxley and Mr. Darwin are agreed that it is not producible. Flourens, and Cuvier, Buffon, and De Candolle, Müller, and John Hunter, Lyell, and Lawrence, Agassiz, and Pouchet, though they know nothing of

[1] Huxley's "Evidence as to Man's Place in Nature." Williams & Norgate : 1863. Pp. 105, 106.
[2] *Ibid.*, p. 108.

the' transmutations hypothecated by Mr. Darwin, yet they do know the "insurmountable barrier" that "Nature" has erected against the change of species.[1] They know that the Linnæan maxim—*Species naturæ opus*—rests on foundations too broad and deep to be shaken by casual excess o, statement, or semblance of perplexity. While mere *varieties*, as superficial excursions from type are technically termed, are never mutually infertile, animals of different species are physiologically contrasted with such varieties by reciprocal repugnance or punitive sterility. The mastiff and the terrier freely inter-breed; not so the horse and the ass: the mongrel dog is a parent; the hybrid mule is not. And the hybrid individual perishes,—*at genus immortale manet*. For it is a fundamental axiom that *animals incapable of common off-spring cannot have sprung from common ancestors*.

On this head therefore, the evidence against Mr. Darwin's theory of the Origin of Species is overwhelming; and no one knows this better than Mr. Huxley himself. When therefore he tells us that he adopts Mr. Darwin's hypothesis, " subject to the production of proof that physiological species *may be* produced by selective

[1] *Vide* Appendix, Note C.

breeding," we are to understand by this that he does not adopt it at all. For, as he is careful to add, "Our acceptance of the Darwinian hypothesis must be provisional so long as one link in the chain of evidence is wanting; and so long as all the animals and plants certainly produced by selective breeding from a common stock are fertile with one another, that link will be wanting."[1]

So long then as Nature remains what it is, "that link,"—Mr. Huxley himself being witness, —will still be "wanting." And yet Mr. Darwin can say—"If a single link in this chain had never existed"! According to Mr. Darwin, Man is what he is, because he has been inextricably linked with the lower animals—with the "ascidian," with the "primordial form" by a chain of which no "single link" is wanting. According to Mr. Huxley, Man is what he is, notwithstanding the chasms in Mr. Darwin's imaginary chain: chasms which Mr. Darwin cannot cross except by "steps" imaginary and aerial, which it is "not difficult to conceive"; but still "steps" which have no corresponding "links" in the world of physiology and fact; steps which cannot be taken at all—not even in imagination —without reversing the Constitution and Course

[1] " Man's Place in Nature," p. 107.

of Nature. For Nature knows nothing of "a group of animals having all the characters exhibited by species" having "ever been originated by selection, whether artificial or natural,"[1]

But although such groups are utterly unknown to Nature they are absolutely necessary to the theory of Ascensive Development. Since therefore they cannot be found, they must be "conceived"; and to conceive them is, in Mr. Darwin's opinion, "not difficult": ("*Facilis descensus*"!) and Prof. Haeckel has conceived them accordingly. Again and again he tells us that *Monera*, worms, and fishes, were "our ancestors." We are reminded that when we speak of "poor worms," or "miserable worms," we should remember that "WITHOUT ANY DOUBT a long series of extinct worms were our direct ancestors."[2] He recognizes twenty-two distinct stages in our evolution; eight of which belong to the invertebrate, and fourteen to the vertebrate sub-kingdom.

Not however until he reaches the Sixth of these imaginary stages does he arrive at the earliest worms, the *Archelminthes*, now represented by the *Turbellaria*. In order to arrive at these "earliest worms," he hypothecates as

[1] " Lay Sermons," p. 295.
[2] " Anthropogenie," p. 399.

number Five, the *Gastræa* (Urdarmthiere) a class of animals purely imaginary. They are placed here because required as ancestors for the *Gastrula*, itself an imaginary order, derived from embryological exigencies.

No. 8 is another imaginary type, called by Haeckel *Chordonia*, because they "developed themselves from the *Annelidæ*, by the formation of a spinal marrow and a *chorda dorsalis*"![1] It is well known that between the *Invertebrata* and the *Vertebrata* there is no transition form. It is also known (by Mr. Darwin) that, by means of the Ascidians, we are supposed to "have at last gained a clue to the source whence the Vertebrata have been derived."[2] But as to that "group of marine animals resembling the larvæ of existing Ascidians," which were our "most ancient progenitors in the kingdom of the Vertebrata":[3]—who they were, or what, or whence, is known to no one but Professor Haeckel! True, even he does not profess to have any producible evidence that such animals ever existed; they are destitute of any single living representative; there is no fossil evidence of their former existence; their sole *raison d'être*

[1] "Natürliche Schöpfungsgeschichte," p. 583.
[2] "Descent of Man," vol. i. p. 205.
[3] *Ibid.*

is, that they are required by the hypothesis. In Haeckel's *Stammbaum* here they are accordingly—as veritable as Falstaff's men in buckram—with no extinct or living representatives, but being, for all that, "undoubtedly" the progenitors of all the Vertebrata, through the Ascidians. Not that they were always so, however. Far from it. But—anticipating the advice of Mrs. Louisa Chick—they knew how much depended on them, and they "made an effort."[1] It succeeded beyond all expectation. They "*developed* THEMSELVES"! How? By the simplest possible process, in the easiest possible manner. Nothing more than—"the formation of a spinal marrow and a *chorda* dorsalis"!

(14) The *Sozura*, is an order of Amphibia interpolated "because required as a necessary transition stage between the true Amphibia," (13,) and (15) The *Protamniota*, or general stem of the mammalia, reptiles, and birds. "What the *Protamniota* were like," says Prof. Huxley, "I do not suppose any one is in a position to say."[2] And yet we are told that

[1] "It's necessary for you to make an effort, and perhaps a very great and painful effort which you are not disposed to make; but this is a world of effort you know, Fanny, and we must never yield, when so much depends upon us. Come! Try!"—*Dombey & Son*, ch. i.

[2] "Critiques and Addresses," p. 318.

"the *Protamniota* split up into two stems, one that of the *Mammalia*, the other common to *Reptilia* and *Aves*."[1] And they are "proved" to have existed (—although no one knows what they were like—) because they were the necessary precursors of

(16), The *Pro-mammalia*, the earliest progenitors of all the Mammalia. And these were followed by (17,) *Marsupialia*, or Kangaroos. "But," says Prof. Huxley, "the existing Opossums and Kangaroos are certainly extremely modified and remote from their ancestors the '*Prodidelphia*,' of which we have not, at present, the slightest knowledge. The mode of origin of the *Monodelphia* from these is a very difficult problem, for the most part left open by Professor Haeckel."[2] Observe: Of these *Prodidelphia* "we have not, at present, the slightest knowledge." And yet this knowledge we "certainly" have: First, that they are the "ancestors" of "the existing Opossums and Kangaroos"; and Second, that these Opossums and Kangaroos "are *certainly* extremely modified and remote from their ancestors the *Prodidelphia*." No wonder that "the mode of origin of the *Monodelphia* from these is a very difficult

[1] "Critiques and Addresses," p. 317.
[2] *Ibid.*, p. 318.

problem." No wonder either that though "the phylum of the *Vertebrata* is the most interesting of all, and is admirably discussed by Prof. Haeckel,"[1] still it certainly does include "a few points which seem," even to Prof. Huxley, "to be open to discussion."[2]

And now we have reached the beginning of the end. For (18) are the *Prosimiæ*, or half-apes, as the indris and loris. And from these, through (19,) the *Menocerca*, or tailed apes, we reach, at last, (20,) the *Anthropoides*, or man-like apes, represented by the modern orang, gibbon, gorilla, and chimpanzee. Not amongst these however are we to look for "the direct ancestors of man, but amongst the *unknown* extinct apes of the Miocene." The Pithecanthropi (21), or dumb ape-men, come next; an unknown race—the nearest modern representatives of which are cretins and idiots.[3] They *must have* existed, in order to furnish means of transition to the final stage (thus far!) *i.e.*, (22) the *Homines*, or true men, who "developed themselves" from their imaginary fathers of the preceding class, "by a gradual conversion of brute howlings into articulate speech."

[1] "Critiques and Addresses," p. 317.
[2] *Ibid.*
[3] "Natürliche Schöpfungsgeschichte," p. 59a.

Apes. 219

Thus then, at last, we reach the goal :—
"There was an Ape."[1] There "*must have been*,"
or there could not have been a man.[2] The
exigency is urgent, and the affirmation easy. It
is only when we proceed to particulars that
difficulties present themselves. Where was this
Ape? And when? And what? No man can
tell.

Haeckel emphatically protests against the
notion that the modern anthropoid apes can be
regarded as our direct progenitors. "Our ape-
like ancestors," he says, "are long since extinct.
Perchance their fossil remains may some time
be found in the tertiary deposits of Southern
Asia or Africa. They must nevertheless be
ranked amongst the tailless catarhine anthropoid apes."[3]

Mr. Darwin includes Europe in the field which
has been so vainly searched for this missing
link. "It is probable," he tells us, "that Africa
was formerly inhabited by extinct apes, closely
allied to the gorilla and chimpanzee; and as
these two species are now man's nearest allies,
it is somewhat more probable that our early
progenitors lived on the African continent than

---
[1] *Vide* Appendix, Note D.
[2] *Ibid.*, Note E.
[3] " Natürliche Schöpfungsgeschichte," p. 577.

elsewhere. But it is useless to speculate on this subject, for an ape nearly as large as a man . . . existed in Europe during the Upper Miocene period ; and since so remote a period the earth has certainly undergone many great revolutions, and there has been ample time for migration on the largest scale."[1] Man's progenitors therefore, like this ape, may have been Europeans, yet Europe, no less than Africa or Asia, has hitherto utterly failed to furnish any fossil remains, either of the immediate, or of the remote, progenitors of man.

"The fossil remains of man hitherto discovered," says Prof. Huxley, "do not seem to me to take us appreciably nearer to that lower pithecoid form, by the modification of which he has, probably, become what he is. . . . Where then must we look for primeval man ? Was the oldest *Homo sapiens* pliocene, or miocene, or yet more ancient ? In still older strata do the fossilized bones of an ape more anthropoid, or a man more pithecoid than any yet known await the researches of some unborn palæontologist ? Time will show."

So be it : *dies declarabit.* But, meantime this doctrine of man's derivation from an unknown ape, in an undiscovered continent, rests—by the

[1] "Descent of Man," vol. i. p. 199.

admissions of its advocates—not on knowledge, but on the want of knowledge. Absolutely powerless to derive man from the ape, it is not less powerless to derive the cardinal ape from the primordial form. And yet it is in the name of Science that we are presented with this paraded pedantry of Nescience, and are asked to believe that "IN THE DIM OBSCURITY of the Past, *we can* SEE"[1] the unreal nonentities, the airy nothings, required by the "theoretic conception," as they "must have" existed—"once upon a time" ![2]

[1] "Descent of Man," vol. ii. p. 389.
[2] *Vide* Appendix, Note F.

CHAPTER XII.

MEN.

"THE question of questions for mankind," says Prof. Huxley, "the problem which underlies all others, and is more deeply interesting than any other—is the ascertainment of the place which Man occupies in nature, and of his relations to the universe of things."[1] For the most part indeed, men are unreflecting as well as uninquiring; "But in every age, one or two restless spirits, blessed with that constructive genius which can only build on a secure foundation,"[2] have adopted sound principles, and proceeded by sure methods, such as those which have now led the Professor to perceive that "though the quaint forms of Centaurs and Satyrs have an existence only in the realms of art, creatures approaching man more nearly than they in essential structure, and yet as thoroughly brutal as the goat's or horse's half of the

[1] "Evidence as to Man's Place in Nature," p. 57.
[2] *Ibid.*

Scientific Sophisms. 223

mythical compound, are now not only known but notorious."[1]

Of these "creatures approaching man in essential structure," yet "thoroughly brutal," the gorilla was once supposed to be the chief. But the day of De Chaillu is over; "because, in my opinion, so long as his narrative remains in its present state of unexplained and apparently inexplicable confusion, it has no claim to original authority respecting any subject whatsoever. It may be truth, but it is not evidence."[2]

The comforting opinion that we had, as men, a cerebral distinction, is also now (alas!) no more. For we are now assured by Prof. Huxley, in direct contradiction to the reiterated declarations of Prof. Owen, that "so far from the posterior lobe, the posterior cornu, and the hippocampus minor being structures peculiar to and characteristic of man, as they have been over and over again asserted to be, even after the publication of the clearest demonstration of the reverse, it is precisely these structures which are the most marked cerebral characters common to man with the apes. They are amongst the most distinctly Simian peculiarities which

[1] "Man's Place in Nature," p. 1.
[2] Ibid., p. 54.

the human organism exhibits." Thus, then, it appears that while Owen and Huxley differ, apes and men do not. It is an unfortunate circumstance that the more we are developed from apes, the more we differ from each other.

But are we then "developed from apes" after all? Is this so certain? This "question of questions for mankind"—how shall we answer it? Shall we accept the dictum of Prof. Huxley, and say that "man is in substance and in structure one with the brutes"? Or shall we pronounce that dictum a mere "theoretic conception," "unverified by observation and experiment"? In either case, what are the facts?

1. And first, as to cerebral *structure*.

"It is clear," says Prof. Huxley, "that man differs less from the chimpanzee or the orang than these do even from the monkeys; and that the difference between the brains of the chimpanzee and of man is almost insignificant, when compared with that between the chimpanzee brain and that of a lemur."

2. As to cerebral *weight*, however, on the other hand, "It must not be overlooked that there is a very striking difference in absolute mass and weight between the lowest human brain and that of the highest ape, a difference which is all the more remarkable when we recol-

lect that a full-grown Gorilla is probably pretty nearly twice as heavy as a Bosjesman, or as many an European woman." "It may be doubted," adds the Professor, "whether a healthy human adult brain ever weighed less than 31 or 32 ounces, or that the heaviest Gorilla brain has exceeded 20 ounces."[1]

3. "This is a very noteworthy circumstance, and doubtless will one day help to furnish an explanation of the great gulf which intervenes between the lowest man and the highest ape in intellectual power, but it has little systematic value" [Why?] "for the simple reason that . . . Regarded systematically, the cerebral differences of man and apes are not of more than generic value, his Family distinction resting chiefly on his dentition, his pelvis, and his lower limbs."[2]

4. On this latter topic, however, Mr. Huxley had previously said, "The pelvis, or bony girdle of the hips, of man is a strikingly human part of his organization."[3] Adding, "But now let us turn to a nobler and more characteristic organ— that by which the human frame seems to be, and indeed is, so strangely distinguished from

[1] "Man's Place in Nature," p. 102.
[2] Ibid., p. 103.
[3] Ibid., p. 76.

all others—I mean the skull." And then, after giving the cubical capacity of the smallest human cranium, and of "the most capacious Gorilla skull yet measured," he says, "Let us assume, for simplicity's sake, that the lowest man's skull has twice the capacity of that of the highest Gorilla."[1]

5. The sum of the statements already quoted, then, is this :—The "Family distinction" of the genus *Homo* is to be found not in his higher, but in his lower, qualities; "resting chiefly," not on the size of his skull, nor on the weight of his brain, but "on his dentition, his pelvis, and his lower limbs." And yet, notwithstanding this,

6. "That by which the human frame is so strongly distinguished from all others" is not the baser structure, but the nobler substance; not his lower limbs, but "a nobler and more characteristic organ . . . the skull."

7. Prof. Huxley need not think it strange if, in despair of reconciling the conflicting members of this duplex thesis—that Man's "family distinction" is not cranial, and yet that by which he is "so strongly distinguished from all others" *is* cranial; that "the great gulf in intellectual power which intervenes between the lowest man and the highest ape" is of little moment, and

[1] "Man's Place in Nature," p. 77.

yet that the organ which indicates that gulf is his "nobler and more characteristic organ;— some readers should relegate it to that category in which he himself has placed a dictum of Prof. Owen's, characterizing it as a "quâ-quâ-versal proposition . . . which may be read backwards, forwards, or sideways, with exactly the same amount of signification."[1]

8. But "quâ-quâ versal" as it is, it does not stand alone. For after we have learned that even when regarded on the lowest grounds, "the pelvis, or bony girdle of the hips, of man is a *strikingly human* part of his organization," and that his Brain is strikingly human in a much higher degree, since it is his Brain, and not his pelvis, which is "to furnish an explanation of the great gulf which intervenes between the lowest man and the highest ape in intellectual power;" we are told—as if to neutralize this concurrent testimony from "structure" and from "substance,"—that "the difference in weight of brain between the highest and lowest man is far greater, both relatively and absolutely, than that between the lowest man and the highest ape." And, in a word, "whatever system of organs be studied, the comparison of their modifications in the ape series leads to one

[1] Man's Place in Nature," p. 106.

and the same result—that the structural differences which separate man from the gorilla and the chimpanzee, are not so great as those which separate the gorilla from the lower apes."

9. Even this latest dictum, if it had been allowed to stand alone, would have been so far definite as to redeem it from the character of "quâ-quâ-versal." But it is not thus allowed. No sooner has it been submissively accepted; no sooner have we brought ourselves with due docility to admit that "the structural differences between man and even the highest apes are small and insignificant," than Prof. Huxley protests he has been misunderstood. "Let me take this opportunity then," says he, "of distinctly asserting, on the contrary, that they are great and significant; that every bone of a Gorilla bears marks by which it might be distinguished from the corresponding bone of a man; and that in the present creation, at any rate, no intermediate link bridges over the gap between *Homo* and *Troglodytes*."[1]

10. This would be conclusive, if only it were final. But it is not final. It is neutralized in the next sentence but one:—" Remember, if you will, that there is no existing link between man and the gorilla; but do not forget that there is a

[1] "Man's Place in Nature," p. 104.

no less sharp line of demarcation, a no less complete absence of any transitional form, between the gorilla and the orang, or the orang and the gibbon. I say not less sharp, though it is somewhat narrower."[1]

11. Can anything be plainer? Prof. Huxley anticipates the result. "On all sides I shall, hear the cry—'We are men and women, not a mere better sort of apes, a little longer in the legs more compact in the foot, and bigger in brain than your brutal chimpanzees and gorillas. The power of knowledge, the conscience of good and evil, the pitiful tenderness of human affections, raise us out of all real fellowship with the brutes, however closely they may seem to approximate us.'" And what is his answer to the objurgation he thus anticipates?

Here it is:—"I have endeavoured to show that no absolute structural line of demarcation, wider than that between the animals which immediately succeed us in the scale, can be drawn between the animal world and ourselves, and I may add the expression of my belief that the attempt to draw a psychical distinction is equally futile, and that even the highest faculties of feeling and of intellect begin to germinate in lower forms of life."[2]

[1] "Man's Place in Nature." [2] *Ibid.*, p. 109.

12. Add to this the further declaration that "our reverence for the ability of manhood will not be lessened by the knowledge that man is, in substance and in structure, one with the brutes."[1] And then contrast with both the words that follow. First, there is no physical distinction: "no absolute structural line of demarcation." Second, there is no psychical distinction: for "the attempt to draw a psychical distinction is equally futile." And third, "even the highest faculties of feeling and of intellect begin to germinate in lower forms of life." And yet, the very next sentence is in these words:—

13. "At the same time no one is more strongly convinced than I am of the vastness of the gulf between civilized man and the brutes: or is more certain that whether *from* them or not, he is assuredly not *of* them."[2]

To harmonize discordant and conflicting assertions like these would be not merely to reconcile the irreconcilable; it would be to show that opposites are identical. Yet until that is done, what else can we say of them but that which their author has already said so wittily of his opponents? They are merely "quâ-quâ-versal propositions . . . which may be

[1] "Man's Place in Nature," p. 112.
[2] *Ibid.*, p. 110.

read backwards, forwards, or sideways, with exactly the same amount of signification."

14. We revert then to our first enquiry: What are the facts? Prof. Huxley's facts are opposed to his conclusions. When he has admitted that between the lowest man and the highest ape there is a general, a particular, and a wide distinction; a distinction which has left its marks on "every bone"; he then proceeds to lay great stress on the fact that, between one family of man and another the difference is greater than between the lowest man and the highest ape."[1] But when he has done this, he proceeds in each case to show that there is a far greater difference between this same ape, and the apes of some other remaining class. But these two statements furnish the important corollary that "there is the same, or an analogous kind of distinction between one family of man and another, and between one family of ape and another." The idea thus suggested is subversive of his theory: viz., that the families of men are sprung from one type, and the families of apes from another; in other words, there is a generic as well as a specific difference between men and apes."

15. Prof. Huxley apart, it is allowed on all

[1] "Man's Place in Nature," p. 78.

hands that socially, morally, religiously, and historically, men and apes are generically distinct. But this distinction as matter of fact either involves a generic distinction between the physiological structure of men and apes, or it does not. If it does, then Mr. Huxley's theory is disproved by the fact; and man is *not* " in substance and in structure one with the brutes." If it does not, then "the cause of this distinction must be looked for elsewhere, and science will have to admit that in man there is an immaterial element which physiology cannot grasp," an element adequate to his elevation at a height so immeasurably above the rest of the animal world.

16. Nor is it to be forgotten that, even by Prof. Huxley himself, this elevation of man above the ape is regarded comparatively as being not merely "immeasurable," but "practically infinite." "Believing as I do, with Cuvier," he says, "that the possession of articulate speech is the grand distinctive character of man," . . . "the primary cause of the UNMEASURABLE and *practically infinite* divergence of the Human from the Simian Stirps." [2]

By universal consent then, nothing is more

[1] "Man's Place in Nature," p. 103 *n.*
[2] *Ibid.*, p. 103 *n.*

certain than that Man is chiefly characterized by those psychical distinctions which in such treatises as that of Prof. Huxley's now cited, are either left entirely out of view, or dismissed in a passing sentence. "Conscience, remorse, ambition, sense of responsibility, improvableness of reason, immense advances in knowledge, self-cultivation, æsthetical sensibilities—these and other qualities of the *Homo sapiens*, not to speak of religious sentiments, broadly and plainly distinguish man from all the Simians and Troglodytes. Grant, for a moment, (what is manifestly inconsistent with the previous statement, that 'the structural differences between man and the highest apes are great and significant') that man is one in substance and structure with these creatures; grant even that their instincts simulate our reason in some remarkable instances; and when all is granted, the vast and varied differences just intimated remain as towering distinctions. To these is added that gift of articulate speech which, though mechanically organized, imparts supreme value to them all; which makes man a communicative being; which gives to a lecturer, such as Professor Huxley, that power to instruct, amuse and illustrate, by which he is raised immeasurably above the cleverest ape that ever climbed a tree, or built a nest, or buried his dead companion under the dried leaves of an African forest."¹

17. As to the alleged ancestry of Man from the brutes, this, then, is certain : "that whether *from* them or not, he is assuredly not *of* them."

But is he "*from* them"? He who answers this question in the affirmative affirms what he cannot even pretend to prove. The evidence, such as it is, in every particular, and in the most positive terms, endorses the direct negative of the proposition which on any theory of Ascensive Development it is found necessary to maintain. It is Mr. Darwin himself who tells us of "the great break in the organic chain between man and his nearest allies, which *cannot be bridged over* by any extinct or living species."² "The fossil remains of man hitherto discovered," says Professor Huxley, "do not seem to take us appreciably nearer to that lower pithecoid form" from which it is conjectured—but only conjectured—that he sprang. It is nothing less than the utter destitution of evidence in support of the unverified "theoretic conception" that constrains even Professor Huxley to ask, "Where then must we look for primæval man?"

[1] *The Athenæum*, No. 1844, p. 288.
[2] "The Descent of Man," vol. i. p. 200.

18. "In the first place, it is manifest that man, the apes, and the half-apes cannot be arranged in a single ascending series, of which man is the term and culmination.

"We may indeed, by selecting one organ or one set of parts, and confining our attention to it, arrange the different forms in a more or less simple manner. But if all the organs be taken into account, the cross relations and inter-dependencies become in the highest degree complex and difficult to unravel."[1] This indeed is generally admitted, but still the theory propounded by Mr. Darwin, and widely accepted, is that "the resemblances between man and apes are such that man *may be conceived* to have descended from some ancient members of the broad-breastboned group of apes," and of all existing apes, the gorilla is regarded as standing towards him in closer relationship than any other.

But what evidence of common origin is afforded by community of structure? "The human structural characters are shared by so many and such diverse forms, that it is impossible to arrange even groups of genera in a single ascending series from the aye-aye to man

[1] "Lessons from Nature," p. 174. By Prof. Mivart. (Murray, 1876.)

(to say nothing of so arranging the several single genera), if all the structural resemblances are taken into account.

"If the number of wrist-bones be deemed a special mark of affinity between the gorilla, chimpanzee, and man, why are we not to consider it also a special mark of affinity between the indris and man? That it should be so considered, however, would be deemed an absurdity by every evolutionist.

"If the proportions of the arms speak in favour of the chimpanzee, why do not the proportions of the legs serve to promote the rank of the gibbons?

"If the obliquely-ridged teeth of Simia and Troglodytes point to community of origin, how can we deny a similar community of origin, as thus estimated, to the howling monkeys and galagos?

"The liver of the gibbons proclaims them almost human; that of the gorilla declares him comparatively brutal.

"The ear-lobule of the gorilla makes him our cousin; but his tongue is eloquent in his own dispraise.

"The slender loris from amidst the half-apes, can put in many a claim to be our shadow refracted, as it were, through a lemurine prism.

"The lower American apes meet us with what seems 'the front of Jove himself,' compared with the gigantic, but low-browed denizens of tropical Western Africa.

"In fact, in the words of the illustrious Dutch naturalists, Messrs. Schrœder, Van der Kolk, and Vrolik, the lines of affinity existing between different Primates construct rather a network than a ladder.

"It is indeed a tangled web, the meshes of which no naturalist has as yet unravelled by the aid of natural selection. Nay, more, these complex affinities form such a net for the use of the teleological retiarius as it would be difficult for his Lucretian antagonist to evade, even with the countless turns and doublings of Darwinian evolutions."[1]

And yet we are told by Professor Tyndall[2] that the naturalist whose mind is "most deeply stored with the choicest materials of the teleologist," rejects teleology. Does he then effect his escape from the reticulations of the complex affinities now specified? By no means. But he refers the spontaneous and independent appearance of these similar structures to "atavism," and "reversion;" to the appearance

[1] Professor Mivart, *ut sup.*, pp. 174, 175.
[2] "Belfast Address."

that is, in modern descendants, of ancient and sometimes long-lost structural characters, which are supposed to have formerly existed in ancestors more or less remote, and wholly hypothetical.

But if this were true: "if man and the orang are diverging descendants of a creature with certain cerebral characters, then that remote ancestor must also have had the wrist of the chimpanzee, the voice of a long-armed ape, the blade-bone of the gorilla, the chin of the siamang, the skull-dome of an American ape, the ischium of a slender loris, the whiskers and beard of a saki, the liver and stomach of the gibbons, and the number of other characters in which the various several forms of higher or lower Primates respectively approximate to man.

"But to assert this is as much as to say that low down in the scale of Primates was an ancestral form so like man that it might well be called an *homunculus;* and we have the virtual pre-existence of man's body supposed, in order to account for the actual first appearance of that body as we know it :—a supposition manifestly absurd if put forward as an explanation."[1]

19. Nor is it an insignificant circumstance,

[1] " Lessons from Nature," p. 176.

as indicating the wholly hypothetical character of the ape ancestry thus assigned to man, that neither on the earth nor under the earth is any trace of such an ancestry discoverable. The number is not small of those who prefer to search the record of the rocks for "Vestiges" of Creation rather than for Footprints of the Creator; but no vestige of man's ascent from the ape is yet producible. In default therefore of evidence adducible from that which is, we are liberally supplied with asseverations as to that which might, or "must have been."

There must, for example, have been "a series of forms graduating insensibly from some ape-like creature to man as he now exists."[1] Now of the series thus alleged, every single member was *ex hypothesi* superior to the lower forms from which he sprang. And Mr. Darwin's doctrine affirms "the survival of the fittest." But while the half-apes are with us to this day the half-men are nowhere. The ape-mothers that found themselves, in the last term of the series, strangely producing men, have perished; while the monkeys, unequal to the production even of apes, have survived. According to the hypothesis the fittest should survive; according to the facts the fittest have perished.

[1] "Descent of Man," vol. i. p. 235.

But this is not all. Besides this imaginary "series of forms," the theory requires further a process of "graduating insensibly." And of this process there is not only no proof, but the evidence, such as it is, points in the direction of disproof. It is Mr. Darwin himself who says, "Breaks incessantly occur in all parts of the series, some being wide, sharp, and defined, others less so in various degrees; as between the orang and its nearest allies—between the Tarsius and the other Lemuridæ." The "intellectual figment" is in evil case when it postulates a process of graduation so gradual as to be imperceptible, yet so abrupt as to exhibit "breaks" which "incessantly occur in all parts of the series," not excluding even "breaks" which are "wide, sharp, and defined." And yet, across these "breaks," Mr. Darwin's theory, by Mr. Darwin's ingenuity, is made to swing its ponderous bulk with an adroit dexterity that might have been envied, in the depths of his African forest, by the ancestral Gorilla himself:—

"All these breaks depend merely on the number of related forms that have become extinct."[1] Could anything be more simple? The "breaks" are there indeed: but they are

[1] "Descent of Man," vol. i. pp. 200, 201.

there only in the absence of the "related forms" "graduating insensibly." You have only to imagine the "forms" and the "breaks" will disappear.

And yet, of these same "forms" it is all the while most certain that they cannot be described; they are not known to have existed; they are not known to have been "related"; they are not known to "have become extinct." Nor are the "breaks" more real. They are breaks only on the assumption of the hypothesis: not otherwise. And the second assumption has no power to confer validity on the first.

20. From this tissue of assumptions we revert to the facts. No less a writer than Mr. Wallace, "the independent originator and by far the best expounder of the theory of Natural Selection," differs *toto cœlo* from Mr. Darwin on the question of the Origin of Man. For the creation of man, as he is, Mr. Wallace postulates the necessity of the intervention of an external Will. He observes that even the lowest types of savages are in possession of capacities far beyond any use to which they can apply them in their present condition, and therefore they could not have been evolved from the mere necessities

of their environment. These capacities have respect to future possibilities of culture. But prolepsis, anticipation, involves intention and a will.

He contends further,[1]—that even as to his body, Man is a clear and palpable and positive exception to the theory of Evolution. To produce the human frame required, he says, the intervention of some special agency. He adverts to the peculiar disposition of the hair on man, especially that nakedness of the back which is common to all races of men, and to the peculiar construction of the hand and foot. "The hand of man," he tells us, "contains latent capacities and powers which are unused by savages, and must have been even less used by palæolithic man and his still ruder predecessors. It has all the appearance of an organ prepared for the use of civilized man, and one which was required to render civilization possible."

Again: speaking of the "wonderful power, range, flexibility, and sweetness of the musical sounds producible by the human larynx," he adds, "the habits of savages give no indication of how this faculty could have been developed." . . . "The singing of savages is a more or less monotonous howling, and the females

[1] "Natural Selection," pp. 332-360.

seldom sing at all." "It seems as if the organ had been prepared in anticipation of the future progress of man, since it contains latent capacities which are useless to him in his earlier condition."[1]

Mr. Wallace is in perfect agreement also with christian theism in the value he attaches to man's "capacity to form ideal conceptions of space and time, of eternity and infinity—the capacity for intense artistic feelings of pleasure, in form, colour, and composition—and those abstract notions of form and number which render geometry possible," as well as with respect to the non-bestial origin of moral perception."[2]

And beyond all this, he considers Man as not only placed "apart, as the head and culminating point of the grand series of organic nature, but as in some degree a new and distinct order of being." . . . "When the first rude spear was formed to assist in the chase; when fire was first used to cook his food; when the first

---

[1] On this subject, indeed, even Mr. Darwin himself admits that "neither the enjoyment nor the capacity of producing musical notes are faculties of the least direct use to man in reference to his ordinary habits of life; they must be ranked amongst the most mysterious with which he is endowed."—*Descent of Man*, vol. ii. p. 333.

[2] "Natural Selection," pp. 351, 352.

seed was sown or shoot planted, a grand revolution was effected in nature, a revolution which in all the previous ages of the earth's history has had no parallel, for a being had arisen who was no longer necessarily subject to change with the changing universe, a being who was in some degree superior to nature, inasmuch as he knew how to control and regulate her action, and could keep himself in harmony with her, not by a change in body, but by an advance in mind."

Against facts like these, of what avail are Mr. Darwin's ingenious speculations? The answer may be given in the words of Professor Mivart. It is the same high authority that pronounced Mr. Darwin's "Origin of Species" to be "a puerile hypothesis," and its distinctive characteristic, "a conception utterly irrational;"[1] who now adds,

"Thus, then, in our judgment the author of the 'Descent of Man' has UTTERLY FAILED in the only part of his work which is really important: . . . and if Mr. Darwin's failure should lead to an increase of philosophic culture on the part of physicists, we may therein find some consolation for the injurious effects which his work is likely to produce on too many of our half-educated classes."[2]

[1] "Lessons from Nature," p. 300. [2] *Ibid.*, p. 184.

Nor is this all. Man is something more than an intellectual animal. He is a free moral agent : and, as such,—and with the infinite future which that freedom opens out before him —he differs from all the rest of the visible universe by "a distinction so profound that no one of those which separate other visible beings is comparable with it. The gulf which lies between his being as a whole, and that of the highest brute, marks off vastly more than a mere kingdom of material beings, and man, so considered, differs far more from an elephant or a gorilla than do these from the dust of the earth on which they tread."[1]

[1] "Lessons from Nature," p. 184.

CHAPTER XIII.

ANIMA MUNDI.

"There lives and works
A Soul in all things : and that Soul is God."
—*Cowper.*

THIS witness is true: and its truth is not impaired by the ignorant positiveness of Agnosticism, or by the positive ignorance of Atheism. As to Atheism, indeed, the verdict already pronounced, after a most minute and searching investigation, is found to be unalterable:—"Every part of the universe is an argument against atheism as a theory thereof."[1] Agnosticism, despite its pretensions to Knowledge, as its very name imports, is a mere confession of Ignorance. And even that ignorance, confronted by the facts of the universe, ceases to be possible when its votaries are willing to cease to be "*willingly* ignorant."

Is there, or is there not, "a Soul in all

[1] Theodore Parker : *Theism, Atheism, and the Popular Theology*, p. 10.

Scientific Sophisms. 247

things?" Theism affirms, Atheism denies, Agnosticism ignores, the existence of any such Soul. To put an end to controversy the appeal is made to facts. Is the affirmation of Theism unsustained by evidence? Is the negation of Atheism consistent with the admissions which Atheism itself has been compelled to make? Is the ignorance of Agnosticism compatible with the knowledge to which Agnosticism makes such arrogant pretensions?

1. "In all things." Let us begin at the beginning. It is in the phenomena of crystallization that Professor Tyndall finds the foundation of all higher phenomena—life, growth, reproduction, intelligence, will. He believes "that the formation of a crystal, a plant, or an animal, is a purely mechanical problem, which differs from the problems of ordinary mechanics in the smallness of the masses and the complexity of the process involved."[1]

Take now the least complex of the three instances of constructive power here mentioned, —that of crystallization. "The human mind," says the Professor, "is as little disposed to look unquestioning at these pyramidal salt crystals

[1] "Fragments of Science," p. 119.

as to look at the pyramids of Egypt without enquiring whence they came.

"How then are those salt pyramids built up? Guided by analogy, you may, if you like, suppose that, swarming among the constituent molecules of the salt, there is an invisible population, controlled and coerced by some invisible master, and placing the atomic blocks in their position. This however is not the scientific idea, nor do I think your good sense would accept it as a likely one. The scientific idea is, that the molecules act upon each other without the intervention of slave labour; that they attract each other and repel each other at certain definite points or poles, and in certain definite directions, and that the pyramidal form is the result of this play of attraction and repulsion. While then the blocks of Egypt were laid down by a power external to themselves, these molecular blocks of salt are self-posited, being fixed in their places by the forces with which they act upon each other."[1]

On this very pertinent analogy it is to be remarked that Professor Tyndall has specified only the points on which it holds good; and here his opponents are in perfect accord with himself. The only point in respect of which they differ from him is that which he has omitted to notice; and in that point the analogy entirely fails.

When, for the slave-labour employed in the construction of the pyramids, we have sub-

[1] "Fragments of Science," pp. 114, 115.

stituted the mutual attractions and repulsions which determine the position of the minute blocks employed in the construction of a crystal, we have dealt with only one element of the problem. That slave-labour was employed not otherwise than as the consequent of antecedent design. The huge blocks of granite or of limestone were not deposited in their relative positions except as those positions had been antecedently determined "by some invisible master;"—the architect who planned; the monarch who ordained and controlled. And when we have said that the infinitesimally minute molecular blocks in a crystal of salt or of sugar are "self-posited," we have indeed dispensed with the necessity of external physical force necessary to simple super-position; but we have made no advance whatever towards a substitute for that Intellectual Force which is (at least) equally necessary in order to symmetrical super-position. We have dispensed with "the intervention of slave-labour." We have not dispensed with the intervention of the Will by which that labour was employed, or the Intelligence by which it was directed, or the Power by which it was controlled. We have dismissed the slaves only. The "Invisible Master" still remains behind.

2. With regard to the pyramids of Egypt all are agreed. Who planned them?

"Was Cheops or Cephrenes architect
Of either Pyramid that bears his name?"

By what agencies were they erected? With what object were they designed? These questions Professor Tyndall regards—and rightly regards—as at once instinctive and inevitable. But when these same questions are put with regard to the "pyramidal salt crystals," whose exquisite finish transcends all architectural composition, the only answer is, that the questions are all at once and altogether out of place.

And yet it is Professor Tyndall who tells us that the very same constitution of mind which compels us to question the pyramids compels us also to question the crystals. Only, the three questions which were inevitable in the former case must, in the latter, be reduced to one. "Who planned?" and "With what object?" are questions inseparable from intelligence in the one case. But in the other, we are told that these are questions with which intelligence has nothing to do. "The scientific idea" is limited exclusively to the one remaining question—the question least interesting and least important of the three—What forces, and what laws operated in their construction?

That the final form of the pyramid expresses the thought of the "invisible master," whether Cheops or Cephrenes, is, on all hands, admitted. How then can it be denied—as it is denied—that the crystal expresses the thought of any intelligence whatever?

3. But, in the crystal, "the molecular blocks are *self*-posited." And, in like manner, the valves of a steam-engine are said to be "*self*-adjusting." But the self-adjustment of the valves, like the self-positing of the molecules, must ultimately be referred to Mind. Except as the result of the operations of a designing mind, there are no valves "self-adjusting," and no molecular blocks "self-posited."

Are we asked to dispense with Mind, because "the agencies by which the crystals are built, are incomparably superior to the agencies employed in the building of the pyramids"? To take this ground is to assume that the more exquisite the agency employed, the less manifest, or the less certain, are the evidences of the operation of mind :—an assumption directly contrary to the fact.

4. The human mind, as Professor Tyndall himself describes it, refuses to rest satisfied with

a reference to "the play of atoms and molecules under the operation of laws." The obvious question instinctively recurs :—How come these atoms and molecules to act with preconcerted harmony, and "like disciplined squadrons under a governing eye, arranging themselves into battalions, gathering round distinct centres, and forming themselves into solid masses,"[1] move with unerring precision towards a predetermined goal? This is the question which, not in consequence of its experience but in virtue of its constitution, the human mind is compelled to ask. To answer it by referring to laws self-constituted, or atoms self-posited, or molecules self-adjusted, is to leave untouched the very thing to be accounted for. What the mind demands a reason for is, the exquisite adjustment here alleged: "and this reason is not rendered by referring the inquirer to the operation of laws; for, apart from and outside of matter, there are no such entities in existence as the laws of matter. The laws of matter are simply the mode, in which matter in virtue of its constitution, acts. Oxygen unites chemically with hydrogen, in certain proportions, under certain conditions, simply because of the qualities or attributes wherewith these two gases are

[1] " Fragments of Science," p. 448.

invested. *It is not the law which determines the combination, but the qualities which determine the law.* These elements act as they act, simply because they are what they are."[1] How then came they to be what they are? These "myriad types of the same letter"; these unhewn blocks from an unknown quarry; more indestructible than adamant; the substratum of all the phenomena of the universe; and yet, amid the wreck of all things else, this infinitude of discrete atoms alone is found incapable of change or of decay. Who preserves to them their absolute identity, notwithstanding their infinite variety? Who endowed them with their inalienable properties? Who impressed upon them the ineffaceable characters which they are found to bear? At what mint were they struck, on what anvil were they forged, in what loom were they woven, so as to possess "all the characteristics of manufactured articles"?

5. Whatever then may be said about "the formation of" "a plant, or an animal," it is certain that the formation of an Atom—and consequently of a crystal—is precisely the opposite of that alleged by Professor Tyndall:—it is *not* "a purely mechanical problem." "Manu-

[1] "Atomism." By Prof. Watts. Belfast: Mullan, 1874, p. 15.

factured articles" may, or may not, be produced by machines; but machines are a product of Mind. And where there is no Mind, there are no "manufactured articles."

6. Between the curiosities of crystallography and the mysteries of life there yawns a gulf measurable only by the whole diameter of being. It is even Haeckel himself who admits that "The phenomena which living things present have no parallel in the mineral world."[1] And yet Professor Tyndall puts the properties of minerals, of mammoths, and of men, into one and the same category; tells us that however strikingly they may be differentiated by specific characters, yet, in every case, this difference is one not of kind, but merely of degree; and that "the formation" of a man, or an oak, equally with that of a snowflake, is nothing more than "a *purely mechanical* problem, which differs from the problems of ordinary mechanics" —not by the introduction of a new element, not by the mysterious origination of vital or mental force,—but only by "the smallness of the masses and the complexity of the process involved."

Now this assertion is not only unsupported by evidence: the evidence completely disproves it.

[1] "History of Creation," voL i. p. 681.

Scientific Sophisms. 255

The points involved in it are two :—First, the introduction of Life. Second, the manifestations of Mind. As to the former of these, Professor Huxley himself declares that—

7. "The present state of knowledge furnishes us with no link between the living and the not-living."[1] Professor Haeckel admits that there is nothing in chemistry that can produce life. That chemistry cannot bridge the colossal chasm between the living and the not-living. That it cannot explain how inorganic is transmuted into organic matter. That "most naturalists, even at the present day, are inclined to give up the attempt at natural explanation 'of the origin of life,' and to take refuge in the miracle of inconceivable creation."[2] In the words of one of them, "We have given up the idea that we can make things grow." Or, to take but one instance more,—the final sentence of Du Bois Reymond,—"It is futile to attempt by chemistry to bridge the chasm between the living and the not-living."

8. Futile as is the attempt however, Professor Huxley has shown himself equal to it. In his most deliberate utterance he tells us that—

[1] Encycl. Brit., Art. "Biology."
[2] "History of Creation," vol. i. p. 327.

"A mass of living protoplasm is simply a molecular machine of great complexity, the total results of the working of which, or its vital phenomena, depend, on the one hand, on its construction, and on the other, upon the energy supplied to it ; and to speak of 'vitality' as anything but the name of a series of operations, is as if one should talk of the horology of a clock."[1]

This oracular deliverance is worthy of the most careful consideration, not less from its own merits than from the celebrity of its author. From it we learn that a "living" thing is "a machine ; " "*simply*" a machine. "The results of the working of" this machine—Milton's "Paradise Lost," for example ; or Shakspere's Plays ; Galileo and Kepler, Newton and Pascal, Socrates and Savonarola, Stephenson and Edison, Turner and Ruskin, — "the *total* results"—are due to two sources. The first of these is "its construction ; " the second, is "the energy supplied to it."

Since, however, to our instructor not less than to ourselves, the "construction" of "a mass of living protoplasm" is an unfathomable secret, of which, notwithstanding his high attainments, even he is profoundly ignorant ; and since "the energy supplied to it" remains now, as ever, an absolutely unknown quantity ; it might

[1] Prof. Huxley, Encyc. Brit., Art. " Biology," 1875.

perhaps have been more candid, as it would certainly have been less misleading, if it had been said at once, and without ambiguous circumlocution, that "its vital phenomena depend" on something of which nothing is known.

It is Prof. Huxley himself who tells us that the "lifeless compounds" carbonic acid, water, and ammonia, cannot combine—cannot, by any wit of man, be combined—so as to "give rise to the still more complex body, protoplasm," unless a principle of life presides over the operation. Unless under those auspices the combination never takes place. But when we ask, What *is* that principle of life? What *is* that presiding Power? We are told that there is no such thing; that "vitality" has no more real existence than "horologity;" and that we might as well speak of a "steam-engine principle," a "watch-principle," or a "railroad-principle," as of a "vital principle," or vital force.

And yet, not even the scathing sarcasm of which Prof. Huxley is a master, can avail to conceal the fact that the analogies thus suggested fail in every particular. The power of a steam-engine is in no degree dependent on its connection with some antecedent steam-engine.

The perfection of a watch is not derived by contact from some other watch. But the perfection of vital movement, and the power of vital force *are* derived by contact, *are* dependent on connection with other, and pre-existing living bodies. Mr. Huxley tells us of something which he finds it convenient to call by the name of "subtle influences." And these "subtle influences," he says, " will convert the dead protoplasm into the living protoplasm;" will "raise the complex substance of dead protoplasm to the higher power, as one may say, of living protoplasm."[1] What are these "subtle influences?" What else are they but vital force?

It is easy to talk of a living body as "a molecular machine," and to attribute "vital phenomena" to its "construction." But what of The Constructor? It is easy to talk of "lifeless compounds" as the "constituents" of a living body. But then these lifeless compounds are "constituents" that do not constitute. They do not even constitute "The Physical Basis of Life." Still less do they constitute the energy of Life itself. "Let the matter be disguised or slurred over as it may, the fact remains that we are utterly unable to imitate vital affinity so far

[1] *Fortnightly Review* for 1869, p. 138.

as to make a bit of material ready for its use, or even to make any definite substance that would have similar chemical relations."[1]

Let it however be supposed, that Prof. Huxley's vaticination has been realized. Let it be assumed that some day "by the advance of molecular physics" the learned Professor will be able to show us how it is that the properties peculiar to water have resulted from the properties peculiar to the gases whose junction constitutes water; and similarly, how the characteristic properties of protoplasm have sprung from properties in the water, ammonia, and carbonic acid that have united to form protoplasm; even then, knowing all this, we should be as far as ever from the more recondite knowledge up to which it is expected to lead. For this knowledge leaves us as ignorant as before concerning that "supplied energy" of Life, without which no protoplasm is ever formed. "To extract the genesis of life from any data that completest acquaintance with the stages and processes of protoplasmic growth can furnish, is a truly hopeless problem. Given the plan of a house, with samples of its brick and mortar, to find the name and nationality of the householder, would be child's play in

[1] Dr. Elam, "Automatism and Evolution."

comparison. Life, as we have seen,[1] is not the offspring of protoplasm, but something which has been superinduced upon, and may be separated from the protoplasm that serves as its material basis. It is therefore distinct from the matter which it animates, and, being thus immaterial, cannot possibly become better known by any analysis of matter."[2]

9. "In every living thing there are *physico-chemical actions*, which also occur out of the body, and *vital actions*. These last, however, are *peculiar to living beings*, and cannot be imitated. In galvanic batteries, and in other arrangements made by man, we may have physico-chemical actions, but *never anything at all like vital actions.*" The physicist "seems to think that pabulum goes into a living thing and becomes changed chemically, just as it may be changed in his laboratory, and the results of this change are work, and certain compounds which are got rid of. In all this, *the living matter* which is absolutely essential in every one of these changes—*without which not one of them could occur*, or even be conceived as occurring in thought, is persistently ignored." "But although the new schools hold it absurd to

[1] *Vide ante*, pp. 119, 120.
[2] Thornton, "Old Fashioned Ethics," pp. 168 *et seq.*

suppose that any peculiar power acting from within or from without can influence the changes in matter, or direct its forces, they see no impropriety in attributing to matter itself, and to force, guiding, and directing, and forming agencies." They transfer to the non-living those active, controlling, and directing powers which have hitherto been regarded as the attributes of life alone. According to them, it is not "will," or "mind," or even "vital force"—it is merely "the inorganic molecule"—that arranges, governs, guides, controls.

Thus, for example, Prof. Huxley has affirmed that a "particle of jelly" *guides* forces. To his mind, he tells us, it is a fact of the profoundest significance that " this particle of jelly is capable of guiding physical forces in such a manner as to give rise to those exquisite and almost mathematically arranged structures, etc."[1] It is not easy to see, however, why the idea of physical forces being guided by a particle of jelly should be accepted as a fact of "profound significance," while the idea of "vitality" acting upon the particles of this jelly, and guiding them and their forces, should be denounced as a fiction, absurd, ridiculous, frivolous, fanciful.

[1] Introduction to the Classification of Animals.

Besides: that physical forces guide matter, is a doctrine neither new nor strange; but here we have the doctrine that matter guides physical forces—a doctrine not less strange than new. "But is it not more probable that neither matter nor force is capable of guiding or directing force or matter? Matter may be said to rule and guide itself, but it can hardly be *ruled and guided by itsel* It might, however, be ruled and guided by something else.

"Concerning the dictum about jelly guiding physical forces, I shall, therefore, venture to remark—1. That living matter is not jelly; 2. That neither jelly nor *matter* is capable of *guiding* or *directing* forces of any kind; and 3. . That the capacity of jelly to guide forces, which Prof. Huxley says is a *fact* of the profoundest significance to him, is not a *fact* at all, but merely an assertion."[1]

10. "If a machine that moved itself could, of itself, divide into new machines, and each take up particles of brass and iron and steel, or other substances entering into its construction, and deposit these in the proper places, so that the several wheels and other elementary parts of the mechanism should grow evenly and regularly, and continue to work while all these

[1] Dr. Beale's "Protoplasm," pp. 74, 75, 77, 81.

changes were proceeding,—such a machine, it is true, would in some particulars be like a living organism." But how stands the fact? "If any apparatus we could contrive developed all possible modes of force—motion, heat, light, electricity, magnetism, chemical action, and any number of others yet to be discovered—that apparatus would still present *no approach whatever* to any organism known. Of course such a thing might be *called* an organism, just as a watch, or a steam-engine, or water, or anything else, may be called a creature,—a worm or any other living thing *called* a machine. But every living machine seems to grow of itself, builds itself up, and multiplies, while every non-living machine that has yet been discovered *is made. It neither grows, nor can it produce machines like itself.*" "Will mechanics account for the movements of an amœba? Where is the being that grows by mechanics, and where is the mechanical apparatus that can be said to grow? Has mechanics taught us the difference between a living seed and the same seed when it has ceased to live?"[1]

11. To revert, for a moment, from "vitality" to "horology." When Mr. Lewes—a writer distinguished for his opposition to what he

[1] Dr. Beale, "Protoplasm," pp. 47, 486.

calls Theological explanations in Science—tells us that we may just as well speak of a watch as the abode of a "watch-force," as speak of the organization of an animal as the abode of a "vital Force,"[1] he is guilty of an oversight common to all those who share his views. It is quite true that the Forces by which a watch moves are natural Forces. But it is the relation of interdependence in which these Forces are placed to each other, or, in other words, the adjustment of them to a particular Purpose, which constitutes the "watchforce;" and the seat of this Force—which is in fact no one Force but a combination of many Forces — is in *the Intelligence which conceived that combination*, and in *the Will which gave it effect*.

"The mechanisms devised by Man are in this respect only an image of the more perfect mechanism of Nature, in which the same principle of Adjustment is always the highest result which Science can ascertain or recognise. There is this difference, indeed,—that in regard to our works we see that our knowledge of natural laws is very imperfect, and our control over them is very feeble; whereas in the machinery of Nature there is evidence of complete knowledge and of absolute control. The universal rule is, that everything is brought about by way of Natural Consequence. But another rule is, that all natural

---

[1] Lewes's "Philosophy of Aristotle," p. 37.

Scientific Sophisms. 265

consequences meet and fit into each other in endless circles of Harmony and of Purpose. And this can only be explained by the fact that what we call Natural Consequence is always the conjoint effect of an infinite number of elementary Forces, whose action and reaction are under direction of the Will which we see obeyed, and of the Purposes which we see actually attained."[1]

12. The relation which an organic structure bears to its purpose in Nature is not less capable of certain recognition than the same relation between a machine and its purpose in human art. "It is absurd to maintain, for example, that the purpose of the cellular arrangement of material in combining lightness with strength, is a purpose legitimately cognisable by Science in the Menai Bridge, but is not as legitimately cognisable when it is seen in Nature, actually serving the same use. The little Barnacles which crust the rocks at low tide, and which to live there at all must be able to resist the surf, have the building of their shells constructed strictly with reference to this necessity. It is a structure all hollowed and chambered on the plan which engineers have so lately discovered as an arrangement of material by which the power of resisting strain or pressure is multiplied in an extraordinary degree. That shell

[1] The Duke of Argyll's "Reign of Law" (Sixth Edition), pp. 124 *et seq.*

is as pure a bit of mechanics as the bridge; both being structures in which the same arrangement is adapted to the same end."[1]

> "Small, but a work divine;
> Frail, but of force to withstand,
> Year upon year, the shock
> Of cataract seas that snap
> The three-decker's oaken spine."[2]

This is but one instance out of a number that no man can count.

The Electric Ray, or Torpedo, has been provided with a Battery which, while it closely resembles, yet in the beauty and compactness of its structure, it greatly exceeds the Batteries by which Man has now learned to make the laws of Electricity subservient to his will. In this Battery there are no less than 940 hexagonal columns, like those of a bees' comb, and each of these is subdivided by a series of horizontal plates, which appear to be analogous to the plates of the Voltaic Pile. The whole is supplied with an enormous amount of nervous matter, four great branches of which are as large as the animal's spinal cord, and these spread out in a multitude of thread-like filaments round the prismatic columns, and finally pass into all

[1] "The Reign of Law," pp. 99, 100.
[2] "Maud."

the cells.[1] "*A complete knowledge of all the mysteries which have been gradually unfolded from the days of Galvani to those of Faraday, and of many others which are still inscrutable to us, is exhibited in this structure.*"

Well may Mr. Darwin say, "It is impossible to conceive by what steps these wondrous organs have been produced."[2] "We see the Purpose—that a special apparatus should be prepared, and we see that it is effected by the production of the machine required: but we have not the remotest notion of the means employed. Yet we can see so much as this, that here again, other laws, belonging altogether to another department of Nature—laws of organic growth—are made subservient to a very definite and very peculiar Purpose." The laws appealed to in the accomplishment of this purpose are at once numerous and highly complicated. They are so because the conditions to be satisfied refer not merely to the generation of Electric force in the animal to which it is given, but to its effect on the nervous system of the animals against which it is to be employed, and also to the conducting medium in which

---

[1] Prof. Owen's "Lectures on Comparative Anatomy," vol. ii. (Fishes).

[2] "Origin of Species." First Edition, p. 192.

both are moving. But the fact that these conditions exist, and must be satisfied, is not the ultimate fact, it is not even the main fact which Science apprehends in such phenomena as these. That which is most observable and most certain, is the manner in which these conditions are met. But this, in other words, is simply the subordination of many laws to a difficult and curious Purpose; a Purpose none the less obvious, and a subordination not the less remarkable, because effected through the instrumentality of mechanical contrivance.

"The new-born Kangaroo," says Professor Owen, "is an inch in length, naked, blind, with very rudimental limbs and tail: in one which I examined the morning after the birth, I could discern no act of sucking: it hung, like a germ, from the end of the long nipple, and seemed unable to draw sustenance therefrom by its own efforts. The mother accordingly is provided with a peculiar adaptation of a muscle (cremaster) to the mammary gland, by which she can inject the milk from the nipple into the mouth of the pendulous embryo. Were the larynx of the little creature like that of the parent, the milk might, probably would, enter the windpipe and cause suffocation: but the fœtal larynx is cone-shaped, with the opening at the apex, which projects, as in the whale-tribe, into the back aperture of the nostrils, where it is closely embraced by the muscles of the 'soft palate. The air-passage is thus completely separated from the fauces, and the injected milk passes in a divided stream, on either side the base of the larynx, into the œsophagus. These correlated modifications of maternal and fœtal structures, designed with especial reference to the peculiar conditions of both mother and offspring, afford, as it seems to me, *irrefragable evidence of Creative foresight.*"[1]

"The parts of this apparatus cannot have produced one another; one part is in the mother, another part in the young one; without their harmony they could not be effective; but nothing except design can operate to make them harmonious. They are *intended* to work together; and we cannot resist the conviction of this intention when the facts first come before us."[2]

13. "A prospect-glass or a forceps is an *instrument;* they have each a final cause; that is, they were each made and adjusted for a certain use. The use of the prospect-glass is to assist the eye; the use of the forceps is to assist the hand. The prospect-glass was made the better to see; the forceps, the better to grasp. *The use did not make* these instruments; they were each *made for the use*—which use was foreseen and premeditated in the mind of the maker of them. We say of each of them without a shadow of hesitation: IF THIS HAD NOT FIRST BEEN A THOUGHT, IT COULD NEVER HAVE BEEN A THING. Now, is the Eye or the Hand an instrument adjusted to a certain use, and thus revealing an antecedent purpose in

[1] *Philosophical Transactions*, 1834, *Reade Lecture*, p. 29.
[2] *Phil. Inductive Sciences*, vol. i. p. 625.

the Creative Mind, or is it not? Can we account for either except by saying that it was thought out before it was wrought out; that it was a concept in mind ere it could possibly appear as a configuration in matter; that before it became *a fact in nature* it must needs have been *a thought in God?*"[1]

14. Can we say that although the prospect-glass is the product of mind, yet no mind presided over the structure of the eye? According to Mr. Darwin, we can and ought. And yet Mr. Darwin begins by admitting it to be apparently "in the highest degree absurd to suppose that *the eye, with all its* INIMITABLE *contrivances* for adjusting the focus to different distances, for admitting different amounts of light, and for the correction of spherical and chromatic aberration, could have been formed by natural selection." He then proceeds to indicate some "probable" stages in the process by which, as he believes, the eye was formed— a process of natural selection, and of that alone. His first postulate is, a nerve specially endowed with sensibility to light. The optic nerve thus—not formed, but—fancied merely, surrounded by pigment cells, and covered by translucent skin, will, in millions of ages, *select*

---

[1] "The Three Barriers," pp. 61 *et seq.*

itself into an eye. Let it be granted :—" in the highest degree absurd " though it be. But the primary postulate—how does Mr. Darwin get that? "How a nerve comes to be sensitive to light," he says, "hardly concerns us more than how life itself originated." Perhaps not: but both questions are studiously evaded when we are left to infer that the nerve made itself, and that life caused itself to live; or, in other words, that both are examples of what Mr. Darwin strangely calls "*variation causing alterations.*"

Take now the several steps of the process as pursued by Natural Selection according to Mr. Darwin; and let but the power competent to do the things which he assumes are done, be credited with sense enough to be aware of its competence, and it may then be regarded as not unlikely to have done some of them on purpose. Whereupon the genesis of the eye ceases to be a mystery. "All the appearances of contrivance that have resulted from the operation find their obvious and complete explanation in the assumption of a contriver, and all such hazy films as that of variability producing variation cease to be capable of serving as excuses for wilful blindness. And why should not the power in question be so credited? Here is Mr. Darwin's solitary reason why. He doubts whether the inference implied may not be 'presumptuous.' He apprehends that we have no 'right to assume that the Creator works by intellectual powers like those of a man.' Truly, of all suggested modes of marking respect for creative power, that of assuming it to have worked un-intelligently is the most original."[1]

"From what I know, through my own speciality, both geometry and experiment, of the structure of lenses and the human eye, I do not believe that any amount of evolution, extending through any amount of time consistent with the requirements of our astronomical knowledge, could have issued in the production of that most beautiful and complicated instrument, the human eye. There are too many curved surfaces, too many distances, too many densities of the media, each essential to the other, too great a facility of ruin by slight disarrangement, to admit of anything short of the intervention of an intelligent Will at some stage of the evolutionary process. The most perfect, and at the same time the most difficult optical contrivance known is the powerful achromatic object glass of a microscope; its structure is the long-unhoped-for result of the ingenuity of

[1] Thornton: "Old-Fashioned Ethics," pp. 238, 239.

many powerful minds; yet in complexity and in perfection it falls infinitely below the structure of the eye. Disarrange any one of the curvatures of the many surfaces, or distances, or densities of the latter; or worse, disarrange its incomprehensible self-adaptive power, the like of which is possessed by the handiwork of nothing human, and all the opticians in the world could not tell you what is the correlative alteration necessary to repair it, and still less to improve it, as natural selection is presumed to imply.'

15. The case is too strong to be explained away. Nature is full of plan, and yet she plans not: she is only plastic to a plan. That plan carries with it its own unanswerable attestation to all healthy understandings. It has its warp indeed, as well as its woof. The exquisite variety of creative adjustments reposes on a basis of fundamental order: exhaustless specialities of adaptation are engrafted on a pervasive unity of type. Morphology, rightly viewed, is not the negation, but one grand phase of the revelation of plan. Teleology is the other. "It has been by following the lamp of Final Cause, and obeying her beckoning hand, that the

[1] Professor Pritchard's Address at the Brighton Congress (1874).

masters of anatomical and physiological science, from Galen to Cuvier, and from Harvey to Owen, have been guided to their splendid discoveries." But the irrepressible question, *For what?* is naturally followed by the further question, *From Whom?* The measure of the confidence with which Science assumes *a use* is the measure of the confidence with which Religion affirms *an Author.* "He that planted the ear, shall He not hear? Or He that made the eye, shall He not see?" This argument has been esteemed unanswerable, not only by the most masculine reasoners among Christian divines, Barrow and Paley, Chalmers and Whewell: "it has carried conviction, from the time of Socrates to that of Cuvier, to the foremost minds of the human race, and found almost its sole antagonists among spinners of cobwebs and dreamers of dreams. . . . The prints of Divine forethought, and the convictions they engender, are scattered over the face of universal nature, and ploughed into the very subsoil of the human mind."

16. To conclude. Modern Materialism then, as expounded by its ablest advocates, whether under the guise of Positive Agnosticism, or that of Scientific Atheism, has no key to unlock the

mysteries of Being. Propounded as a theory of the Universe, it has no commencement and no continuity. There are "First Beginnings" of which it has no knowledge. There are Barriers which it cannot pass, and chasms which it cannot cross, and deeps which it cannot fathom, and mysteries which it cannot even pretend to explain. That extension which we call space; that duration which we call time; that substance which we call matter; whence came they? "There shall be no Alps"—? They shall be explained away? Matter shall be defined in terms of Mind? Space and Time shall be declared non-entities—non-existent outside the faculties of the Being percipient?[1]

But then whence came this Being? and whence came his faculty "percipient"? Matter, too, however defined, is possessed of certain properties, and constituted in definite proportions, and specified in distinct categories—carbon, gold, iodine, etc.,—whence came all these? Then too, besides material properties, there are material forces. Heat is a mode of motion; and motion is a result of force; and force operates according to law. But who ordained the Law? and who upholds it? Who established "the sequence of events as observed by

[1] See Appendix, Note G.

us"? Who originated Motion? Where is the primal Force?

"We will assume that science has done its utmost; and that every chemical or animal force is demonstrably resolvable into heat or motion, reciprocally changing into each other. I would myself like better, in order of thought, to consider motion as a mode of heat than heat as a mode of motion: still, granting that we have got thus far, we have yet to ask, What is heat? or what motion? What is this 'primo mobile,' this transitional power, in which all things live, and move, and have their being? It is by definition something different from matter, and we may call it as we choose—'first cause,' or 'first light, or 'first heat', but we can show no scientific proof of its not being personal, and coinciding with the ordinary conception of a supporting spirit in all things."[1]

"The Lord of all, Himself through all diffused,
Sustains, and is the life of all that lives.
Nature is but a name for an effect,
Whose cause is God."

With Him is the breath of Life. With Him is the secret of Power. This is what men of science " are finding more and more, below their

[1] "The Queen of the Air:" by John Ruskin, LL.D. (1869), p. 74.

facts, below all phenomena which the scalpel and the microscope can show; a something nameless, invisible, imponderable, yet seemingly omnipresent and omnipotent, retreating before them deeper and deeper, the deeper they delve; that which the old schoolmen called 'forma formativa,' the mystery of that unknown and truly miraculous element in nature which is always escaping them, though they cannot escape it; that of which it was written of old, 'Whither shall I go from Thy presence, or whither shall I flee from Thy Spirit?'"[1]

17. Proof? See it in the great gulf between the organic and the inorganic, the living and the not-living, a grain of sand and a grain of corn. See it in the inscrutable phenomena of growth. See it in the immutable order which dominates the countless varieties of the vegetable world. Amid all those varieties, with their corresponding powers, it does not matter in the least by what concurrences of circumstance or necessity they may gradually have been developed: the concurrence of circumstance is itself the supreme and inexplicable fact. "We always come at last to a formative cause, which directs the circumstance and mode of meeting

[1] Canon Kingsley. Lecture at Sion College.

it. If you ask an ordinary botanist the reason of the form of a leaf, he will tell you it is a 'developed tubercle,' and that its ultimate form 'is owing to the directions of its vascular threads.' But what directs its vascular threads? 'They are seeking for something they want,' he will probably answer. What made them want that? What made them seek for it thus? Seek for it, in five fibres or in three? Seek for it, in serration, or in sweeping curves? Seek for it in servile tendrils, or impetuous spray? Seek for it in woollen wrinkles rough with stings, or in glossy surfaces, green with pure strength, and winterless delight?" It is Mr. Ruskin who asks these questions: and it is Mr. Ruskin who adds, "There is no answer."[1]

Then too this leaf, whatever its form, is alive. It points, not more to a Formative Cause than to a Living Power. Polarity of atoms, molecular movements, chemical affinities, may be adduced to explain, even while in fact they conceal, the phenomena of structure and configuration in the inorganic world. But when the chemical affinities are brought under the influence of the air, and of solar heat, the formative force enters an entirely different phase. "It does not now merely crystallize indefinite

[1] "Queen of the Air," p. 104.

masses, but it gives to limited portions of matter the power of gathering, selectively, other elements proper to them, and binding these elements into their own peculiar and adopted form." But this "power of gathering selectively," the power that catches out of chaos charcoal, water, lime, or what not, and fastens them down into a given form, the power that is continually creating its own shells of definite shape out of the wreck round it,—*What* is it? and Whence? "There is no answer."

Next comes the gap which separates vegetable from animal life. "These are necessarily the converse of each other, the one deoxidizes and accumulates, the other oxidizes and expends. Only in reproduction or decay does the plant simulate the action of the animal, and the animal never, in its simplest forms, assumes the functions of the plant. Those obscure cases in the humbler spheres of animal and vegetable life which have been supposed to show a union of the two kingdoms, disappear on investigation." This is the testimony of Principal Dawson, who adds, "This gap can, I believe, be filled up only by an appeal to our ignorance."[1]

[1] "Story of the Earth and Man." By J. W. Dawson, LL.D., F.R.S., F.G.S., etc. Hodder and Stoughton, 1873, p. 326. [See p. 298, to which this Reference belongs.]

Of the chasms which separate species, the same author writes,—" It was this gap, and this only, which Darwin undertook to fill up by his great work on the origin of species, but notwithstanding the immense amount of material thus expended, it yawns as wide as ever, since it must be admitted that no case has been ascertained in which an individual of one species has transgressed the limits between it and other species." [1]

Transcending all the rest is the gulf that separates the brute from man. It is Professor Huxley himself who tells us that the "divergence of the Human from the Simian Stirps" is "immeasurable and practically infinite." Who made it so? Huxley believes, with Cuvier, that "the possession of articulate speech is the grand distinctive character of man." But whence did he derive an endowment so unique and so invaluable? "Men have *words*, which are projected ideas; brutes have only *sounds*, which are projected sensations. Brutes vociferate: men speak. The physical organization is wedded to the mental capacity—a mouth, and wisdom. Neither, apart, would effloresce into Language: both must conspire and combine. So the one mind which has thoughts to

[1] See Appendix, Note H.

be interpreted is furnished in the human tongue with an all-accomplished interpreter." But whence came this "one mind which has thoughts to be interpreted"?
What is the origin of Mind? What is the genesis of Thought?

18. For Thought is no mere "function of the brain"; nor is it "medullary matter that thinks." "The function of the lung is not unintelligible; it can be followed throughout, and understood throughout. Though the peculiarity of vitality mingles there, it can still, in a certain aspect, be called a physical function, and its result is of an identical nature. If, and so far as, the function is physical, the result is physical. So with the stomach : function and result are there in the same category of being. The liver is so far a physical organ that it can be seen, it can be touched, it can be handled ; but is it otherwise with the bile, which is the result of its function? Can it too, not be seen, and touched, and handled? Is it not essentially of the same nature? Is it not physical, in the same way and to the same extent as the liver is physical? But look now to the brain, and the so-called product of *its* function. Do we any longer find the same identity of the terms? No; the terms there are veritable extremes—

extremes wider than the poles apart—extremes sundered by the whole diameter of being. The result here, then, is not like the result of any other function. *It is wholly unique;* something quite new, fresh, and original; something unprecedented, something unparalleled, absolutely single and singular, absolutely *sui generis.* The result here in fact, is the very antithesis, the very counterpart of the organ which is supposed to function it.

"An organ, after all, consists of parts; but thought has no parts, thought is one. Matter has one set of qualities; Mind, another; and these sets are wholly incommensurable, wholly incommunicable. A feeling is not square, a thought is not oval. . . . No function of the body, and no function of any machine out of the body, presents any parallel to the nature of thought."[1]

Before this problem of the genesis of Thought, Materialism is dumb. And yet this same Thought ("without precedent," "without parallel,") has changed the face of the world. "From the moment when the first skin was used as a covering, when the first rude spear was formed to assist in the chase, the first seed sown

[1] Dr. Stirling's " Materialism in relation to the Study of Medicine," p. 8.

or root planted, a grand revolution was effected in nature, *a revolution which in all the previous ages of the world's history had had no parallel*, for a being had arisen who was no longer necessarily subject to change with the changing universe,— a being who was in some degree superior to nature, inasmuch as he knew how to control and regulate her action, and could keep himself in harmony with her, not by a change in body, but by an advance in mind.

"Here then we see the true grandeur and dignity of man. On this view of his special attributes, we may admit that even those who claim for him a position and an order, a class, or a sub-kingdom by himself, have some reason on their side. He is, indeed, a being apart, since he is not influenced by the great laws which irresistibly modify all other organic beings. . . . Man has not only escaped 'natural selection' himself, but he is actually able to take away some of that power from nature which, before his appearance, she universally exercised."[1]

Conclusive as is this testimony in itself, it is doubly so on account of the quarter from which it comes. From a very different quarter comes

[1] Mr. Wallace, in the *Anthropological Review*, May, 1864.

the characteristic, but concurrent testimony of Thomas Carlyle :—

"Capabilities there were in me" (says Teufelsdröckh) "to give battle, in some small degree, against the great Empire of Darkness: does not the very Ditcher and Delver, with his spade, extinguish many a thistle and puddle; and so leave a little Order, where he found the opposite? Nay, your very Daymoth has capabilities in this kind ; and ever organizes something (into its own Body, if no otherwise), which was before Inorganic ; and of mute dead air makes living music, though only of the faintest, by humming.

"How much more, one whose capabilities are spiritual; who has learned, or begun learning, *the grand thaumaturgic art of Thought!* Thaumaturgic I name it ; for hitherto all Miracles have been wrought thereby, and henceforth innumerable will be wrought ; whereof we, even in these days, witness some. Of the Poets' and Prophets' inspired Message, and how *it makes and unmakes whole worlds*, I shall forbear mention : but cannot the dullest hear Steam-engines clanking around him ?"[1]

What then, is the origin, and who is the originator of "that subtle force which we term Mind"?

19. Man, as defined by Professor Huxley,[2] is "a conscious automaton," "endowed with free-will"; and in his Essay on "The Physical Basis of Life" he confesses that "*our volition counts for*

[1] "Sartor Resartus," chap. iv.
[2] *Fortnightly Review*, November, 1874, p. 577.

something as a condition of the course of events"; and that this "can be verified experimentally as often as we like to try."[1] This machine which is not mechanical; this automaton with a will of its own; this creature whose actions are at once automatic and autonomic; this "automaton endowed with free-will," is a novel invention quite worthy of Mr. Huxley's ingenuity. But whence did it derive the faculties with which he says it is endowed?

It is "conscious," he tells us. And its "volition counts for something." What then is Volition? and whence? And what is Consciousness?

"Can you satisfy the human understanding in its demand for logical continuity between molecular processes and the phenomena of consciousness?" It is Professor Tyndall who asks this question, and his answer to it is this:—

"This is a rock on which materialism must inevitably split whenever it pretends to be a complete philosophy of life."[2]

And with the candid and elegant Lucretian, Professor Huxley—notwithstanding his materialistic declaration of faith in molecular machinery

[1] "Lay Sermons," p. 145.
[2] Belfast Address.

—agrees. "What consciousness is," he says, "we know not; and how it is that anything so remarkable as a state of consciousness comes about as the result of irritating nervous tissue, is just as unaccountable as the appearance of the Djin when Aladdin rubbed his lamp in the story."[1]

"Afferent nerves lie here, and carry to; efferent nerves lie there, and carry from; but in none of them—neither in fibre of nerve nor in fibre of brain, will you find any hint of consciousness. How any material impressions should awake thought; but, still more, how, in independence of all impressions, thought should be all the while there, alive and active, A WORLD BY ITSELF—that is the mystery. And that no scalpel, no microscope, will ever explain. Mechanical balances the most delicate, chemical tests the most sensitive, are all powerless there. And why? Simply because consciousness and they are incommensurable, of another nature, of another world from the first, sundered from each other, as I have said, by the whole diameter of being."[2]

"It is quite true that the tympanum of the ear vibrates under sound, and that the surface of

[1] Huxley's "Physiology," p. 193.
[2] Stirling's "Materialism" *ut sup.*, p. 7.

the water in a ditch vibrates too; but the ditch hears nothing for all that; and my hearing is still to me as blessed a mystery as ever, and the interval between the ditch and me, quite as great. If the trembling sound in my ears was once of the marriage-bell which began my happiness, and is now of the passing-bell which ends it, the difference between those two sounds to me cannot be counted by the number of concussions. There have been some curious speculations lately as to the conveyance of mental consciousness by 'brain-waves.' What does it matter how it is conveyed? The consciousness itself is not a wave. It may be accompanied here or there by any quantity of quivers and shakes, up or down, of anything you can find in the universe that is shakeable—what is that to me? My friend is dead, and my —according to modern views—vibratory sorrow is not one whit less, or less mysterious, to me, than my old quiet one."[1]

Whence came then this emergence of Personal Consciousness among the world of living creatures? From what source have we derived that sense of individual personality which constitutes "an altogether new and original fact, one which cannot be conceived as developed or

[1] Ruskin: 'Athena," p. 70.

developable out o. any pre-existing phenomena or conditions"? That consciousness of an I Myself, or Personality, which asserts an antithesis between the Man, and all that the Man makes his own—whence came it, if not from that Eternal Consciousness, that Divine Personality Who, when He made us, made us in His Own image?

20. Science, in the modern doctrine of the Conservation of Energy, and the Convertibility of Forces, insists, with increasing emphasis, that all kinds of Force are but forms or manifestations of some one Central Force issuing from some one Fountain-head of Power. Sir John Herschel has not hesitated to say, that "it is but reasonable to regard the Force of Gravitation as the direct or indirect result of a Consciousness or a Will existing somewhere."[1] But if for the phenomena of the material world you must have an external Will, how much more for those which characterize the World of Mind! "A will that hangs by the Central Will" is intelligible: but, refuse to recognise that Central Will, and then how can you account for that "lord paramount," the Human Will?

"Two things," said Immanuel Kant, "are awful to me: the starry firmament, and the

[1] "Outlines of Astronomy." Fifth Edition, p. 291.

sense of Responsibility in Man." And again : " Duty! wondrous thought, that workest neither by fond insinuation, flattery, nor by any threat, but merely by holding up thy naked 'law in the soul,' and so extorting for thyself always reverence, if not always obedience; before whom all appetites are dumb, however secretly they rebel ; WHENCE THY ORIGINAL?"

Enough. Nature is a hierarchy, and the head is Man. " Mind, language, civilization, worship—the will to determine, the tongue to speak, the hand to do—these, in their boundless purport, are all awanting till the Creator plants upon the scene the solitary owner of the Perfect Brain. Named in one word, all these are *wisdom;* and Man, ' thinker of God's thoughts after Him,' is, among uncounted myriads of lower existences, on this earth the Only Wise." [1]

" This universe is not an accidental cavity, in which an accidental dust has been accidentally swept into heaps for the accidental evolution of the majestic spectacle of organic and inorganic life. That majestic spectacle is a spectacle as plainly for the eye of reason as any

[1] "The Three Barriers," p. 96.

diagram of mathematic. That majestic spectacle could have been constructed, *was* constructed, only in reason, for reason, and by reason. From beyond Orion and the Pleiades, across the green hem of earth, up to the imperial personality of man, all, the furthest, the deadest, the dustiest, is for fusion in the invisible point of the single Ego—*which alone glorifies it. For* the subject, and on the model of the subject, all is made." [1]

"But the stone doth not deliberate whether it shall descend, nor the wheat take counsel whether or not it shall grow. Even men do not advise how their hearts shall beat, though without that pulse they cannot live. What then can be more clear than that those natural agents which work constantly, *for those ends which they themselves cannot perceive*, must be directed by some high and over-ruling wisdom, and who is that but the great Artificer who works in all of them? . . . For, as 'every house is builded by some man,' and the earth *bears no such creature of itself;* stones do not grow into a wall, or first hew and square, then unite and fasten themselves together; trees sprout not cross like dry and sapless beams, nor spars and tiles

[1] "As Regards Protoplasm," p. 37.

Scientific Sophisms.

arrange themselves into a roof; as these are the supplies of art, and testimonies to the understanding of man, the great artificer on earth, so is the world itself but a house, the habitation and the handiwork of an Infinite Intelligence, and '*He who built all things is* GOD.'"[1]

ᾧ ἡ δόξα εἰς τοὺς αἰῶνας τῶν αἰώνων. ἀμήν.

[1] Pearson: "On the Creed," Art. I. *Vide infrà*, Appendix, Note K.

APPENDIX.

NOTE A. PAGE 5.

CHORUS.

a.

Life and the universe show spontaneity:
Down with ridiculous notions of Deity
Churches and creeds are all lost in the mists:
Truth must be sought with the Positivists.

ζ.

If you are pious (mild form of insanity),
Bow down and worship the mass of Humanity.
Other religions are buried in mists,
We're our own Gods, say the Positivists.

EUELPIDES.

These Positivists are very positive.

PEISTHETAIRUS.

And very negative too. I can't agree
With folk who fancy they're their own creators.

"The British Birds. By the Ghost of Aristophanes," (Mortimer Collins), 1872. P. 47, *et seq.*

NOTE B. PAGE 27.

In the Third Edition of his "First Principles" (Stereotyped), Mr. Spencer, concluding his observations on this topic, says,—"From the remotest past which Science can fathom, up to the novelties of yesterday, an essential trait of Evolution has been the transformation of the homogeneous into the heterogeneous." And his last word on the subject is this :—

"As we now understand it, Evolution is definable as a change from an incoherent homogeneity to a coherent heterogeneity, accompanying the dissipation of motion and integration of matter."—Pp. 359, 360.

NOTE C. PAGES 175 and 214.

Lawrence, who quotes in confirmation the words of Cuvier, thus concludes his disquisition on the subject :— "We may conclude, then, from a general review of the preceding facts, that nature has provided, by the INSURMOUNTABLE BARRIER of instinctive aversion, of sterility in the hybrid offspring, and in the allotment of species to different parts of the earth, against any corruption or change of species in wild animals. We must therefore admit, for all the species which we know at present, as sufficiently distinct and constant, a distinct origin and common date."—*Lectures on Physiology*. First Edition. P. 261.

Cuvier had previously said,—"La nature a soin d'empêcher l'alteration des espèces, qui pourroit résulter de leur mélange, par l'aversion mutuelle qu'elle leur a donnée : il faut toutes les ruses, toute la contrainte de l'homme pour faire contracter ces unions, même aux espèces qui se ressemblent le plus . . . aussi ne voyons nous pas dans nos bois d'individus intermediaires entre le lièvre et le lapin, entre le cerf et le daim, entre la marte et la fouine?"—*Discours Preliminaire.* P. 76. (See also P. 71).

And subsequently, M. Flourens,—"Il y a deux caractères qui font juger de l'espèce : la *forme*, comme dit M. Darwin, ou la *ressemblance*, et le *fécondité*. Mais il y a longtemps que j'ai fait voir que la ressemblance, la forme, n'est qu'un caractère accessoire : *le seul caractère essentiel est la* FÉCONDITE. . . . L'espèce est d'une *fécondité continue*, et toutes les variétés sont entre elles d'une *fécondité continue*, ce qui prouve qu'elles ne sont pas sorties de l'espèce, qu'elles restent espèce qu'elle ne sont que l'espèce, qui s'est diversement nuancée. Au contraire, les espèces sont distinctes entre elles, *par la raison décisive*, qu'il n'y a entre elles qu'une *fécondité bornée*. J'ai déjà dit cela, mais je ne saurais trop le redire."—"*Examen du Livre de M. Darwin, Sur l'Origine*," etc. Pp. 34-36.

NOTE D. PAGE 221.

"There was an APE in the days that were earlier
Centuries passed, and his hair became curlier;
Centuries more gave a thumb to his wrist—
Then he was MAN, and a Positivist."
    ("The British Birds," *ut sup.*, p. 48.)

NOTE E. PAGE 221.

"11. Now these are the generations of the higher vertebrata. In the cosmic period the Unknowable evoluted the bipedal mammalia.

12. And every man of the earth while he was yet a monkey, and the horse while he was a hipparion, and the hipparion before he was an oredon.

13. Out of the ascidian came the amphibian and begat

the pentadactyle; and the pentadactyle by inheritance and selection produced the hylobate, from which are the simiadæ in all their tribes.

14. And out of the simiadæ the lemur prevailed above his fellows, and produced the platyrhine monkey.

15. And the platyrhine begat the catarrhine, and the catarrhine monkey begat the anthropoid ape, and the ape begat the longimanous orang, and the orang begat the chimpanzee, and the chimpanzee evoluted the what-is-it.

16. And the what-is-it went into the land of Nod and took him a wife of the longimanous gibbons. .

17. And in process of the cosmic period were born unto them and their children the anthropomorphic primordial types.

18. The homunculus, the prognathus, the troglodyte, the autochthon, the terragen :—these are the generations of primeval man."—*The New Cosmogony.*

NOTE F. PAGE 223.

"'Will you have why and wherefore, and the fact
Made plain as pikestaff?' modern Science asks.
'That mass man sprung from was a jelly-lump
Once on a time; he kept an after course
Through fish and insect, reptile, bird and beast,
Till he attained to be an ape at last
Or last but one.'"

"Prince Hohenstiel-Schwangau : Saviour of Society." By Robert Browning. Smith, Elder & Co., 1871. P. 68.

NOTE G. PAGE 282.

"Except by neglecting to distinguish between sight and hearing, the effects, and light and sound, their respective causes, it would surely have been impossible for Professor Huxley to come to the strange conclusion that if all living beings were blind and deaf, 'darkness and silence would everywhere reign.' Had he not himself previously explained that light and sound are peculiar motions communicated to the vibrating particles of an universally diffused ether, which motions, on reaching the eye or ear, produce impressions which, after various modifications, result eventually in seeing or hearing? How these motions are communicated to the ether matters not. Only it is indispensable to note that they are not communicated by the percipient owner of the eye or ear, so that the fact of there being no percipient present cannot possibly furnish any reason why the motions should not go on all the same.

"But as long as they did go on there would necessarily be light and sound; for the motions are themselves light and sound. If, on returning to his study in which, an hour before, he had left a candle burning and a clock ticking, Professor Huxley should perceive from the appearance of candle and clock that they had gone on burning and ticking during his absence, would he doubt that they had likewise gone on producing the motions constituting and termed light and sound, notwithstanding that no eyes or ears had been present to see or hear? But if he did not doubt this, how could he any more doubt that, although all sentient creatures suddenly became eyeless and earless, the sun might go on shining, and the wind roaring, and the sea bellowing as before?"
—*Thornton's "Huxleyism."*

NOTE H.   PAGE 287.

It is important to observe that not a few of those who strenuously maintain a doctrine of Evolution, (though not Mr. Darwin's doctrine,) not a few even of Mr. Darwin's

X

most ardent admirers, maintain at the same time and not less strenuously, that the facts in relation to that theory are altogether inexplicable, apart from the recognition of an Intelligent Designer, a presiding Mind, a Universal Power, creative, formative, sustaining.

Thus, for instance, Mr. Thornton, while eulogizing what he calls "the soundness of all the main and really essential principles of Darwinism," exposes with just severity the incompetence and inadequacy of the theories adopted—and necessarily adopted—by those teleologists who reject teleology.

When Mr. Darwin attempts to account for Instinct by hypothecating the accumulation of slight variations from a primordial type—"variations produced by the same unknown causes as those which produce slight deviations of bodily structure,"—Mr. Thornton replies: "But here I am once more compelled to join issue with him. Of the causes which he styles unknown, I maintain that we know at least thus much—either they are themselves intelligent forces, or they are forces acting under intelligent direction; and in support of this proposition I need not perhaps do more than show from Mr. Darwin's example what infinitely harder things must be accepted by those who decline to accept this."

Having done this most elaborately, and conceded the long list of "admissions" for which "not a little liberality is required," he thus concludes:—

"Let us, however, liberally waive this and all similar objections, and assume a community of hive bees to have been, in the *utterly unaccountable* manner indicated by the term spontaneous variation, developed from a meliponish stock. Unfortunately, all our liberality will be found to have been thrown away without perceptibly simplifying the problem to be solved. For whatever be among meliponæ the distribution of the generative capacities, among hive bees, at any rate, *all workers are sterile neuters, which never have any offspring to whom to bequeath their cell-making skill, while the queen-bee and drones, which alone can become parents, have no such skill to be-*

queath. Clearly, the formula of 'descent with modification by natural selection,' is, in its literal sense, utterly inapplicable here. In whatever manner the cell-making faculty might have been acquired by the first homogeneous swarm of hive bees, it must inevitably have terminated with the generation with which it commenced, if transmission by direct descent had been necessary for its continuance. The only resource open to Mr. Darwin is to suppose not merely (what is indeed, obviously the fact) that queen-bee after queen-bee, besides generating each in turn a progeny of workers endowed with instincts which their parents did not possess and could not therefore impart, generated also princess-bees destined in due season to generate a working progeny similarly endowed with instincts underived from their parents; but to suppose, further, that all this has happened in the total absence of aim, object, intention, or design.

"Now that all this should have so happened, although not absolutely inconceivable; nor, therefore, absolutely impossible, is surely too incredible to be believed except in despair of some other hypothesis a trifle less preposterous. It is surely not worth while to set the doctrine of probabilities so completely at naught, for the sake of AN EXPLANATION WHICH AVOWEDLY LEAVES EVERY DIFFICULTY UNEXPLAINED, referring them all to causes not simply unknown but unconjecturable.

"What excuse then have philosophers, of all people, for doing this in preference to the simple expedient of supposing that, *although the parturient bee.* queen o. other, *cannot intend* that any of her progeny should be more bounteously endowed than herself, '*there is* AN INDEPENDENT INTELLIGENCE *that does so intend?*'"—"Recent Phases ot Scientific Atheism.'

NOTE J   PAGE 190.

THE FINE OLD ATOM MOLECULE.

AIR.—" *The Fine Old English Gentleman."*

(*To be sung at all gatherings oj aavanced Sciolists and Scientists.*)

We'll sing you a grand new song, evolved from a 'cute young pate,
Of a fine old Atom-Molecule of prehistoric date,

In size infinitesimal, in potencies though great,
And self-formed for developing at a prodigious rate—
    Like a fine old Atom-Molecule,
    Of the young World's proto-prime!

In it slept all the forces in our cosmos that run rife,
To stir Creation's giants or its microscopic life;
Harmonious in discord, and coöperant in strife,
To this small cell committed, the World lived with his Wife—
    In this fine old Atom-Molecule,
    Of the young World's proto-prime!

In this autoplastic archetype of Protean protein lay
All the humans Space has room for, or for whom Time makes a day,
From the Sage whose words of wisdom Prince or Parliament obey,
To the Parrots who but prattle, and the Asses who but bray—
    So full was this Atom-Molecule,
    Of the young World's proto-prime!

All brute-life, from Lamb to Lion, from the Serpent to the Dove,
All that pains the sense or pleases, all the heart can loathe or love,
All instincts that drag downwards, all desires that upwards move,
Were caged, a "happy family," cheek-by-jowl and hand-in-glove,
    In this fine old Atom-Molecule,
    Of the young World's proto-prime!

In it Order grew from Chaos, Light out of Darkness shined,
Design sprang up by Accident, Law's rule from Hazard blind,

Appendix. 301

The Soul-less Soul evolving—against, not after, kind—
As the Life-less Life developed, and the Mind-less ripened Mind,
  In this fine old Atom-Molecule,
  Of the young World's proto-prime !

Then bow down, Mind, to Matter; from brain-fibre, Will, withdraw;
Fall Man's heart to cell Ascidian, sink Man's hand to Monkey's paw ;
And bend the knee to Protoplast in philosophic awe—
Both Creator and Created, at once work and source of Law,
  And our Lord be the Atom-Molecule,
  Of the young World's proto-prime !

Punch.

NOTE K. PAGE 298.

While these latter sheets are passing through the press, there appears in *The World*, the paragraph here subjoined ; a paragraph interesting and important under any circumstances, but under existing circumstances, doubly so.

"Frank Buckland died on the 19th ultimo [*i.e.*, Dec. 1880], working to the last. Two days before (on the 17th), he finished the preface to his latest book, the *Natural History of British Fishes*. From early sheets of that preface, I make the following extract, in which the dying man—evidently, from the context, not then knowing himself dying—makes a declaration of belief which is wholly antagonistic to the theories of Darwin and his school :—

"'I have another object in writing this book ; it is to endeavour to show the truth of the good old doctrines of the Bridgewater Treatises, which have so ably demonstrated the "*power, wisdom, and goodness of God, as manifested in the Creation.*" Of late years the doctrines of so-called "Evolution" and "Development," have seemingly gained ground amongst those interested in natural history; but I have too much faith in the good sense and natural acumen of my fellow countrymen to think that these tenets will be very long-lived. To put matters very straight, I steadfastly believe that the Great Creator, as indeed we are directly told, made all things perfect and "very good" from the beginning; perfect and very good every created thing is now found to be, and will so continue to the end of time.'"

M.B.B.A.

BOOKS IN
THE STANDARD LIBRARY.
THEIR STERLING WORTH.
OPINIONS OF CRITICS.

I.
Life of Cromwell.

NEW YORK SUN:
"Mr. Hood's biography is a positive boon to the mass of readers, because it presents a more correct view of the great soldier than any of the shorter lives published, whether we compare it with Southey's, Guizot's, or even Forster's."

PACIFIC CHURCHMAN, San Francisco:
"The fairest and most readable of the numerous biographies of Cromwell."

GOOD LITERATURE, New York:
"If all these books will prove as fresh and readable as Hood's 'Cromwell,' the literary merit of the series will be as high as the price is low."

NEW YORK DAILY GRAPHIC:
"Hood's 'Cromwell' is an excellent account of the great Protector. Cromwell was the heroic servant of a sublime cause. A complete sketch of the man and the period."

CHRISTIAN UNION, New York:
"A valuable biography of Cromwell, told with interest in every part and with such condensation and skill in arrangement that prominent events are made clear to all."

SCHOOL JOURNAL, New York:
"Mr. Hood's style is pleasant, clear, and flowing, and he sets forth and holds his own opinion well."

EPISCOPAL RECORDER, Philadelphia:
"An admirable and able Life of Oliver Cromwell, of which we can unhesitatingly speak words of praise."

NEW YORK TELEGRAM:
"Full of the kind of information with which even the well-read like to refresh themselves."

INDIANAPOLIS SENTINEL, Ind.:
"The book is one of deep interest. The style is good, the analysis searching, and will add much to the author's fame as an able biographer."

THE WORKMAN, Pittsburgh, Pa.:
"This book tells the story of Cromwell's life in a captivating way. It reads like a romance. The paper and printing are very attractive."

NEW YORK HERALD:
"The book is one of deep interest. The style is good, the analysis searching."

II.
Science in Short Chapters.

JOURNAL OF EDUCATION, Boston:
"'Science in Short Chapters' supplies a growing want among a large class of busy people, who have not time to consult scientific treatises. Written in clear and simple style. Very interesting and instructive."

ACADEMY, London, England :

"Mr. Williams has presented these scientific subjects to the popular mind with much clearness and force. It may be read with advantage by those without special scientific training."

RELIGIOUS TELESCOPE, Dayton, Ohio :

"It is historic, scientific, and racy. A book of intense practical thought, which one wishes to read carefully and then read again."

NEW YORK SCHOOL JOURNAL:

"A volume of handy science, not only interesting as an abstract subject, but valuable for its clear expositions of every-day science. Of Professor Williams as an authority upon such subjects, it is unnecessary to comment. He already has a fame as a scientific writer which needs no recommendation."

PALL MALL GAZETTE, London, England :

"Original and of scientific value."

GRAPHIC, London :

"Clear, simple, and profitable."

CANADA BAPTIST, Toronto :

"A rich book at a marvellously low price. The style is sprightly and simple. Every chapter contains something we all want to know."

NEWARK DAILY ADVERTISER, N. J.:

"As an educator this book is worth a year's schooling, and it will go where schools of a high grade cannot penetrate. For such a book twenty-five cents seems a ridiculous sum."

J. W. BASHFORD, Auburndale, Mass. :

"A marvellous book, as fascinating as Dickens, to be consulted as an authority along with Britannica, and even fuller of practical hints than the latter's articles. I do not know how you can print its 300 pages for 25 cents."

AMERICAN, Philadelphia :

"Mr. Williams' work is a practical compendium."

III.
The American Humorist.

COMMERCIAL GAZETTE, Cincinnati, Ohio :

"It is finely critical and appreciative ; exceedingly crisp and unusually entertaining from first to last."

CHRISTIAN INTELLIGENCER, New York :

"A book of pleasant reading, with enough sparkle in it to cure any one of the blues."

CONGREGATIONALIST, Boston :

"They are based upon considerable study of these authors, are highly appreciative in tone, and show a perceptivity of American humor which is yet a rarity among Englishmen."

SALEM TIMES, Mass.:

"No writer in England was, in all respects, better qualified to write a book on American Humorists than Haweis."

CHRISTIAN JOURNAL, Toronto :

"We have been specially amused with the chapter on poor Artemus Ward, which we read on a railway journey. We fear our fellow-passengers would think something ailed us, for laugh we did, in spite of all attempts to preserve a sedate appearance."

OCCIDENT, San Francisco :

"This book is pleasant reading, with sparkle enough in it—as the writer is himself a wit—to cure one of the 'blues.' "

DANBURY NEWS, Conn.:

"Mr. Haweis gives a brief bibliographical sketch of each writer mentioned in the book, an analysis of his style, and classifies each into a distinct type from the others. He presents copious extracts from their works, making an entertaining book."

CENTRAL BAPTIST, St. Louis:
"A perusal of this volume will give the reader a more correct idea of the character discussed than he would probably get from reading their biographies. The lecture is analytical, penetrative, terse, incisive, and candid. The book is worth its price, and will amply repay reading."

SCHOOL JOURNAL, New York:
"Terse and brief as the soul of wit itself."

INDIANAPOLIS SENTINEL, Indiana:
"It presents, in fine setting, the wit and wisdom of Washington Irving, Oliver W. Holmes, James R. Lowell, Artemus Ward, Mark Twain, and Bret Harte, and does it con amors."

THE MAIL, Toronto, Ont.:
"Rev. H. R. Haweis is a writer too well-known to need commendation at our hands for, at least, his literary style. The general result is that not a page repels us and not a sentence tires. We find ourselves drawn pleasantly along in just the way we want to go; all our favorite points remembered, all our own pet phrases praised, and the good things of each writer brought forward to refresh one's memory. In fine, the book is a most agreeable companion."

LUTHERAN OBSERVER, Philadelphia:
"The peculiar style, the mental character, and the secret of success, of each of these prominent writers, are presented with great clearness and discrimination."

IV.
Lives of Illustrious Shoemakers.

WESTERN CHRISTIAN ADVOCATE, Cincinnati:
"When we first took up this volume we were surprised that anybody should attempt to make a book with precisely this form and title. But as we read its pages we were far more surprised to find them replete with interest and instruction. It should be sold by the scores of thousands."

PRESBYTERIAN OBSERVER, Baltimore:
"The writer of this book well understands how to write biography—a gift vouchsafed only to a few."

NEW YORK HERALD:
"The sons of St. Crispin have always been noted for independence of thought in politics and in religion; and Mr. Winks has written a very readable account of the lives of the more famous of the craft. The book is quite interesting."

DANBURY NEWS, Conn.:
"The STANDARD LIBRARY has been enriched by this addition."

LITERARY WORLD, London:
"The pages contain a great deal of interesting material—remarkable episodes of experience and history."

BOSTON GLOBE:
"A valuable book, containing much interesting matter and an encouragement to self-help."

CHRISTIAN STANDARD, Cincinnati:
"It will inspire a noble ambition, and may redeem many a life from failure."

CHRISTIAN SECRETARY, Hartford, Conn.:
"Written in a sprightly and popular manner. Full of interest."

EVANGELICAL MESSENGER, Cleveland:
"Everybody can read the book with interest, but the young will be specially profited by its perusal."

LEICESTER CHRONICLE, England:
"A work of the deepest interest and of singular ability."

COMMERCIAL GAZETTE, Cincinnati:
"One of the most popular books published lately."

CENTRAL METHODIST, Kentucky:
"This is a choice work—full of fact and biography. It will be read with interest, more especially by that large class whose awl and hammer provide the human family with soles for their feet."

THE WESTERN MAIL, England:
"Written with taste and tact, in a graceful, easy style. A book most interesting to youth."

CHRISTIAN GUARDIAN, Toronto:
"It is a capital book."

EVANGELICAL CHURCHMAN, Toronto:
"This is a most interesting book, written in a very popular style."

V.

Flotsam and Jetsam.

SATURDAY REVIEW, Eng.:
"Amusing and readable.... Among the successful books of this order must be classed that which Mr. Bowles has recently offered to the public."

NEW YORK WORLD:
"This series of reflections, some philosophic, others practical, and many humorous, make a cheerful and healthful little volume, made the more valuable by its index."

CENTRAL METHODIST, Cattlesburgh, Ky.:
"This is a romance of the sea, and is one of the most readable and enjoyable books of the season."

LUTHERAN OBSERVER, Phil.:
"The cargo on this wreck must have been above all estimate in value. How much 'Jetsam' there may be we cannot tell, but what we have seen is all 'Flotsam,' and will float and find its way in enriching influence to a thousand hearts and homes."

NEW YORK HERALD:
"It is a clever book, full of quaint conceits and deep meditation. There is plenty of entertaining and original thought, and 'Flotsam and Jetsam' is indeed worth reading."

CHRISTIAN ADVOCATE, Nashville, Tenn.:
"Many of the author's comments are quite acute, and their personal tone will give them an additional flavor."

METHODIST RECORDER, Pittsburgh, Pa.:
"In addition to the charming incidents related, it fairly sparkles with fresh and original thoughts which cannot fail to interest and profit."

GOOD LITERATURE, New York:
"... Never fails to amuse and interest, and it is one of the pleasantest features of the book that one may open it at a venture and be sure of finding something original and readable."

HERALD AND PRESBYTER, Cincinnati, Ohio:
"His manner of telling the story of his varied observations and experiences, with his reflections accompanying, is so easy and familiar, as to lend his pages a fascination which renders it almost impossible to lay down the book until it is read to the end."

NEW YORK LEDGER:
"It is quite out of the usual method of books of travel, and will be relished all the more by those who enjoy bits of quiet humor and piquant sketches of men and things on a yachting journey."

NEW YORK STAR:
"Not too profound for entertainment, and yet pleasantly suggestive. A volume of clever sayings."

CHRISTIAN SECRETARY, Hartford, Conn.:
"It is a book well worth reading, ... full of thought."

PRESBYTERIAN JOURNAL,
Philadelphia:
"A racy, original, thoughtful book. On the slight thread of sea-voyaging it hangs the terse thoughts of an original mind on many subjects. The style is so spicy that one reads with interest even when not approving."

CHRISTIAN INTELLIGENCER, New York:
"No one can spend an hour or two in Mr. Bowles' gallery of graphic pen-pictures without being so deeply impressed with their originality of conception and lively, spicy expression, as to talk about them to others."

VI.
The Highways of Literature.

NATIONAL BAPTIST, Phila.:
"A book full of wisdom; exceedingly bright and practical."

PACIFIC CHURCHMAN, San Francisco:
"The best answer we have seen to the common and most puzzling question, 'What shall I read?' Scholarly and beautiful."

DANBURY NEWS:
"Its hints, rules, and directions for reading are, just now, what thousands of people are needing."

CHRISTIAN WITNESS, Newmarket, N. H.:
"Clear, terse, elegant in style. A boon to young students, a pleasure for scholars."

NEW YORK HERALD:
"Mr. David Pryde, the author of 'Highways of Literature; or, What to Read, and How to Read,' is an erudite Scotchman who has taught with much success in Edinburgh. His hints on the best books and the best method of mastering them are valuable, and likely to prove of great practical use."

NEW YORK TABLET:
"This is a most useful and interesting work. It consists of papers in which the author offers rules by which the reader may discover the best books, and be enabled to study them properly."

VII.
Colin Clout's Calendar.

LEEDS MERCURY, England:
"The best specimens of popular scientific expositions that we have ever had the good fortune to fall in with."

NEW YORK NATION:
"The charm of such books is not a little heightened when, as in this case, a few touches of local history, of customs, words, and places are added."

AMERICAN REFORMER, New York:
"There certainly is no deterioration in the quality of the books of the STANDARD LIBRARY. This book consists of short chapters upon natural history, written in an easy, fascinating style, giving rare and valuable information concerning trees, plants, flowers, and animals. Such books should have a wide circulation beyond the list of regular subscribers. Some will criticise the author's inclination to attribute the marvellous things which are found in these plants, animals, etc., to a long process of development rather than to Divine agency. But the information is none the less valuable, whatever may be the process of these developments."

EDINBURGH SCOTSMAN, Scotland:
"There can be no doubt of Grant Allen's competence as a writer on natural history subjects."

NEW YORK HERALD:
"A book that lovers of natural history will read with delight. The author is such a worshipper of nature that he gains our sympathy at once."

THE ACADEMY, London:
"The point in which Mr Grant Allen is beyond rivalry is in his command of language. By this we do not mean only his rich vocabulary, but include also his arrangement of thought and his manipulation of sentences. We could imagine few better lessons to a pupil of English than to be set to analyze and explain the charm of Mr. Grant Allen's style."

CANADIAN BAPTIST, Toronto:
"Mr. Grant Allen is one of the few scientific men who can invest common natural objects and processes with poetical beauty and make them attractive to ordinary readers."

HERALD, Monmouth, Oregon:
"A wonderful book by a charming naturalist. Lovers of flowers, birds, plants, etc., will prize this volume highly."

NEW YORK JOURNAL OF COMMERCE:
"A charming volume, free from the taint of exaggeration and sensationalism."

INDIANAPOLIS SENTINEL:
"He is as keen an observer as Thoreau or Burroughs."

CHRISTIAN STANDARD, Cincinnati:
"They are written in a pleasant and captivating style, and contain much valuable information."

METHODIST PROTESTANT, Baltimore.
"One of the finest productions of modern times."

GOOD LITERATURE, New York
"A trustworthy guide in natural history, as well as a delightful, entertaining writer."

VIII.
George Eliot's Essays.

THE CRITIC, New York:
"Messrs. Funk & Wagnalls have done a real service to George Eliot's innumerable admirers by reprinting in their popular STANDARD LIBRARY the great novelist's occasional contributions to the periodical press."

NEW YORK SUN:
"In the case of George Eliot especially, whose reviews were anonymous, and who could never have supposed that such fugitive ventures would ever be widely associated with the name of a diffident and obscure young woman, we gain access in her early essays, as in no other of her published writings, to the sanctuary of her deepest convictions, and to the intellectual workshop in which literary methods and processes were tested, discarded, or approved, and literary tools fashioned and manipulated long before the author had discerned the large purposes to which they were to be applied. . . . Looking back over the whole ground covered by these admirable papers, we are at no loss to understand why George Eliot should have made it a rule to read no criticisms on her own stories. She had nothing to learn from critics. She was justified in assuming that not one of those who took upon themselves to appraise her achievements had given half of the time or a tithe of the intellect, to the determination of the right aims and processes of the English novel which as these re-

views attest, she had herself expended on that object before venturing upon that form of composition which Fielding termed the modern epic."

EXAMINER, New York:
"These essays ought to be read by any one who would understand this part of George Eliot's career; and, indeed, they furnish the key to all her subsequent literary achievements."

BROOKLYN DAILY EAGLE:
"It is rather suprising that these essays have not been collected and published before, and it is a matter of congratulation that they are now issued."

CHRISTIAN ADVOCATE, New York:
"They show the versatility of the great novelist. One on Evangelical Teaching is especially interesting."

INDIANAPOLIS SENTINEL:
"Nathan Shepherd's introduction to these essays is worth many times the price of the volume."

EPISCOPAL METHODIST, Baltimore:
"Everybody of culture wants to read all George Eliot wrote."

NORTHERN CHRISTIAN ADVOCATE, Syracuse:
"The compiler of this collection has done a unique and a useful work."

METHODIST PROTESTANT, Baltimore:
"They comprise some of the best of the author's writings."

ZION'S HERALD, Boston:
"As remarkable illustrations of her masculine metaphysical ability as is evidenced in her strongest fictions."

CHURCH UNION, New York:
"Nathan Sheppard, the collector of the ten essays in this form, has written a highly laudatory but critical introduction to the book, on her 'Analysis of Motives,' and after reading it, it seems to us that every one who would read her works profitably and truly, should first have read it."

HARTFORD EVENING POST:
"They are admirable pieces of literary workmanship, but they are much more than that. . . . These essays are triumphs of critical analysis combined with epigrammatic pungency, subtle irony, and a wit that never seems strained."

IX.

Charlotte Brontë.

DAILY ADVERTISER, Newark, N. J.:
"There was but one Charlotte Brontë, as there was but one William Shakespeare. To write her life acceptably one must have made it the study of years; have studied it in the integrity of all its relations, and considered it from the broadest as well as from the narrowest aspect. This is what Mrs. Holloway has done."

ZION'S HERALD, Boston:
"This well-written sketch, with selections from her writings, will be appreciated and give a clear idea of the remarkable intellectual ability of this gifted but heavily-burdened woman."

NEW YORK HERALD:
"There are, at times, flights of eloquence that rise to grandeur."

BROOKLYN DAILY EAGLE:
"Managed with the rare skill we might expect at the hands of a fair-minded woman dealing with the traits of character and the actual career of one who, amid extraordinary circumstances of adversity, plodded her way to fame within the span of a brief lifetime."

SOUTHERN CHURCHMAN, Richmond, Va.:
"There are few memoirs more sad than those of this gifted woman and her sisters. An interesting volume at the cheap price of fifteen cents."

JOURNAL AND MESSENGER, Cincinnati:

"The reader, for a small sum, will obtain quite a thorough understanding of the characteristics and literary ability of Miss Brontë, and also be placed in possession of some of her rarest thoughts."

EPISCOPAL RECORDER, Philadelphia:

"The manner in which the reminiscences are narrated is very agreeable, and the reader wonders how so fascinating a life-story could be found in the prosy confines of literature. . . . A thoroughly enjoyable style of description and a deep sympathy with the subject render Mrs. Holloway's sketch exceedingly readable."

CENTRAL CHRISTIAN ADVOCATE, St. Louis:

"The book will be welcomed by all lovers of pure bibliographical literature."

X.
Sam Hobart.

DAILY FREE PRESS, London, Ontario:

"The continual additions made to the STANDARD LIBRARY of works of a high order is evidence that the reading public can easily absorb something more useful than the mere novel. . . . The latest issue deals with the life of a railroad engineer—Sam Hobart, one of the million men who are employed in the railway service of America. . . . It is a marvel of cheapness and biographical excellence."

NEW YORK WORLD:

"A graphic narrative and a strong picture of a life full of heroism and changes. Full of encouragement, and as thrilling as a romance."

GUARDIAN, Truro, Nova Scotia:

"The author's object in writing it was to portray the possibilities of happiness and usefulness within the reach of a workingman content to fill the sphere of usefulness awarded him, and willing to lend a helping hand to do work for God and humanity. It is just such a book as we would like to see in the hands of railroad men."

DANBURY NEWS:

"It is doubtful if any working person can read this book, and not become a better worker and a better man."

EPISCOPAL METHODIST, Baltimore:

"A charming book. All railroad men will be interested in it, and it will pay professional men to read it."

CHRISTIAN SECRETARY, Hartford, Conn.:

"The object of the book is to show how happy and useful a workingman may be, if content in his work and willing to do well. Written in a very interesting way, and while it will probably be devoured by railroad men, it will afford very pleasurable reading to all."

RELIGIOUS HERALD, Hartford, Conn.:

"An entertaining book designed to aid in making one true and noble."

LUTHERAN OBSERVER, Philadelphia:

"Dr. Fulton has done a good work in writing this story of a railroad man. It is a genuine record of heroic fidelity to duty. Let it be scattered by the thousands."

CHURCH ADVOCATE, Harrisburgh:

"If every workingman and employer would follow its principles, the solution of the *Labor Question* would be near at hand."

XI.
Successful Men.

JOURNAL OF EDUCATION, Boston:
"This book possesses all the charm of biography of distinguished men, and abounds in witty, humorous, and telling anecdotes and illustrations."

INTER-OCEAN, Chicago:
"The style is terse, vigorous, and pleasant, abounding in sententious maxims, which are well calculated to impress young readers. Nowhere have we found more incentives to honorable living so delightfully and impressively told than in this volume. If it could be stuffed into every boy's satchel as he journeys from home it would be well."

CHRISTIAN UNION, New York:
"We cordially commend this book to young men."

YOUNG CHURCHMAN, Milwaukee:
"Full of good maxims and sound advice for the young."

BROOKLYN (N.Y.) EAGLE:
"A wonderfully instructive book."

LUTHERAN OBSERVER, Philadelphia:
"Clear, forcible, pungent — nearly every page sparkles with fresh illustration or a pertinent story."

CHRISTIAN SECRETARY, Hartford, Ct.:
"Full of sound, wise, and practical advice to all young men of all occupations. Written with an earnest and noble purpose to help and encourage our youth. It is placed at a low price, and ought to have a wide circulation."

NEW YORK OBSERVER:
"This book will no doubt be found helpful to those who apprehend the truth most easily when presented concretely and in a pictorial form."

DANBURY NEWS, Conn.:
"Invaluable to the youth standing on the threshold of manhood."

PRESBYTERIAN JOURNAL, Philadelphia:
"Clear, stirring, convincing, suggestive, and highly beneficial."

ZION'S HERALD, Boston:
"A capital book to place in the hands of young men commencing a business or professional career."

EVENING CHRONICLE, New Orleans:
"An excellent book of its kind. Pleasant reading. Contains many hints both original and practical."

OCCIDENT, San Francisco:
"Full of sprightly and interesting matter."

CHRISTIAN CHRONICLE, Montpelier Vt.:
"Worth many times its price."

PAGE COURIER, Luray, Va.:
"We would like to see this book in the hands of every youth. Its truths are so forcibly stated that they cannot fail to impress deeply the minds of those who read it. High as we prize our copy, we will loan it to any young man who will promise to read it."

XII.
Nature Studies.

JOURNAL OF COMMERCE, New York:
"One of the freshest of the series. Richard A. Proctor's name is attached to some of the most entertaining papers in the volume."

CHRISTIAN SECRETARY, Hartford, Conn. :

" This volume is replete with interest and general nformation concerning secrets wrested from the tight grasp of nature."

THE CRITIC, New York :

" Were we to act upon the principle that good wine needs no brush, we should certainly forbear praising the 'potable gold' presented in 'Nature Studies.' The twenty-four essays are at once agreeable reading and intellectually stimulative."

DANBURY (Ct.) NEWS:

" Although by a scientist, the book is not a teacher of scepticisms. Proctor believes fully in the existence of an all-creating, all-ruling God. But his views of the Creator are greater than ours, because his knowledge of the vastness of time, of space, and of creation are greater than ours. The book is intensely interesting, as well as thoroughly instructive."

GOOD LITERATURE, New York :

" Mr. Allen writes most delightfully in this volume . . . showing always a most thorough sympathy with nature in all her subtle provisions for the support of her favorite children."

SCHOOL JOURNAL, New York :

" Richard A. Proctor is the editor-in-chief of this volume, and this fact is a guarantee of merit in the contents. . . . There is the greatest variety in subjects, and the reader is sure of genuine intellectual entertainment wherever he opens the book."

METHODIST RECORDER, Pittsburg :

" These eminent naturalists give us in this volume many articles as interesting and as exciting as a story in human life, and there is not one that will disappoint the most dull reader. The theories advanced in some of the articles will probably not be accepted, but will be of interest to show the light in which these theories are held by their advocates."

CHURCH ADVOCATE, Harrisburg, Pa. :

" This is a valuable book."

PRESBYTERIAN JOURNAL, Philadelphia :

" These essays are intensely entertaining and attractive. In reading one should always carefully discriminate between the facts and the hypotheses of the scientists. The volumes of this series are admirable for summer reading."

CHRISTIAN JOURNAL, Toronto, Ont. :

" The essays bristle with indisputable and most interesting facts."

ZION BIBLE TEACHER, Portland, Maine :

" It is worthy of remark when such a work as this can be bought for this price."

PRESBYTERIAN WITNESS, Halifax, N. S. :

" A large amount of valuable reading from five of the greatest scientists of the day."

XIII.
India, What Can it Teach Us?

THE GLOBE, Boston :

" To judge of the value of Funk & Wagnalls' Standard Library, one may take up an issue at random, each one of the series thus far being very choice. But in ' India : What Can it Teach Us ?'

its most scholarly work is published. It contains some of the richest fruits of Max Müller's life-long study of Sanscrit literature. Such an instructive work at so low a price ought to be eagerly seized upon."

www.ingramcontent.com/pod-product-compliance
Lightning Source LLC
Chambersburg PA
CBHW022049230426
43672CB00008B/1115